3D Integration of Resistive Switching Memory

This book offers a thorough exploration of the three-dimensional integration of resistive memory in all aspects, from the materials, devices, array-level issues, and integration structures to its applications.

Resistive random-access memory (RRAM) is one of the most promising candidates for next-generation nonvolatile memory applications owing to its superior characteristics including simple structure, high switching speed, low power consumption, and compatibility with standard complementary metal oxide semiconductor (CMOS) process. To achieve large-scale, high-density integration of RRAM, the 3D cross array is undoubtedly the ideal choice. This book introduces the 3D integration technology of RRAM, and breaks it down into five parts:

1 Associative Problems in Crossbar array and 3D architectures;
2 Selector Devices and Self-Selective Cells;
3 Integration of 3D RRAM;
4 Reliability Issues in 3D RRAM;
5 Applications of 3D RRAM beyond Storage.

The book aspires to provide a relevant reference for students, researchers, engineers, and professionals working with resistive random-access memory or those interested in 3D integration technology in general.

Qing Luo received his Ph.D. from the Institute of Microelectronics, Chinese Academy of Sciences (IMECAS), Beijing, China, in 2017. He is currently Professor at the Key Laboratory of Microelectronics Devices and Integrated Technology in IMECAS. His research interests are emerging memory devices including resistive RAM devices and ferroelectric memory devices.

Frontiers in Semiconductor Technology

Chief Editor: Sheng-Kai Wang
Institute of Microelectronics of Chinese Academy of Sciences, hereinafter IMECAS

Editorial Board Members:
Hong Dong *(Nankai University, China)*
Xuan-Wu Kang *(IMECAS)*
Bo Li *(IMECAS)*
Xiaoyang Lin *(Beihang University, China)*
Xiuyan Li *(Shanghai Jiaotong University, China)*
Hongliang Lu *(Fudan University, China)*
Qing Luo *(IMECAS)*
Qing-Zhu Zhang *(IMECAS)*
Rui Zhang *(Zhejiang University, China)*
Yu Zhao *(Harbin Institute of Technology, China)*

Semiconductor technology has been perhaps the most prominent technology industry in modern society over the past 70 years. Facing the future, emerging technologies are constantly shaping the industry and promoting its continuous development.

Outstanding young scientists from various technology sectors have been invited to join this book series. Through this platform, the aim is for the books within the series to provide new insights and contributions to the development of modern semiconductor technology. The scope of the series is wide, covering semiconductor physics, materials, device processes, equipment, IC design methods amid many other topics while studies involving case studies and applied settings will also be prominent. The titles included in the series are designed to appeal to students, researchers and professionals across semiconductor science and engineering as well as interdisciplinary researchers from many scientific disciplines.

Titles in the series currently include:

Kinetic Studies in GeO2/Ge System
A Retrospective from 2021
Sheng-Kai Wang

Please contact us (wangshengkai@ime.ac.cn, lian.sun@informa.com) if you have an idea for a book for the series.

3D Integration of Resistive Switching Memory

Edited by
Qing Luo

CRC Press is an imprint of the
Taylor & Francis Group, an informa business

First edition published 2023
by CRC Press
6000 Broken Sound Parkway NW, Suite 300, Boca Raton, FL 33487-2742

and by CRC Press
4 Park Square, Milton Park, Abingdon, Oxon, OX14 4RN

CRC Press is an imprint of Taylor & Francis Group, LLC

© 2023 selection and editorial matter, Qing Luo; individual chapters, the contributors

Reasonable efforts have been made to publish reliable data and information, but the author and publisher cannot assume responsibility for the validity of all materials or the consequences of their use. The authors and publishers have attempted to trace the copyright holders of all material reproduced in this publication and apologize to copyright holders if permission to publish in this form has not been obtained. If any copyright material has not been acknowledged please write and let us know so we may rectify in any future reprint.

Except as permitted under U.S. Copyright Law, no part of this book may be reprinted, reproduced, transmitted, or utilized in any form by any electronic, mechanical, or other means, now known or hereafter invented, including photocopying, microfilming, and recording, or in any information storage or retrieval system, without written permission from the publishers.

For permission to photocopy or use material electronically from this work, access www.copyright.com or contact the Copyright Clearance Center, Inc. (CCC), 222 Rosewood Drive, Danvers, MA 01923, 978-750-8400. For works that are not available on CCC please contact mpkbookspermissions@tandf.co.uk

Trademark notice: Product or corporate names may be trademarks or registered trademarks and are used only for identification and explanation without intent to infringe.

ISBN: 978-1-032-48943-8 (hbk)
ISBN: 978-1-032-48950-6 (pbk)
ISBN: 978-1-003-39158-6 (ebk)

DOI: 10.1201/9781003391586

Typeset in Times
by MPS Limited, Dehradun

Contents

Contributors

Qing Luo
Professor at Institute of Microelectronics
Chinese Academy of Sciences
China

Yaxin Ding
Ph.D Candidate at Institute of Microelectronics
Chinese Academy of Sciences
China

Xumeng Zhang
Associate Professor at Fudan University
China

Xiaoxin Xu
Associate Professor at Institute of Microelectronics
Chinese Academy of Sciences
China

Jianguo Yang
Associate Professor at Institute of Microelectronics
Chinese Academy of Sciences
China

Tiancheng Gong
Associate Professor at Institute of Microelectronics
Chinese Academy of Sciences
China

Dengyun Lei
Associate Professor at Guangdong University of Technology
China

Introduction 1

Qing Luo

The development of social networking and the emergence of the Internet of Things (IoT) have led to an explosion in global data volumes. According to a report from Internet Data Center (IDC), as shown in Figure 1.1, the total amount of global data volumes will reach 175 ZB by 2025. However, in the traditional Von-Neumann architecture, there is a "memory wall" between DRAM and NAND Flash, which limits the performance of the entire computer system.

To tear down the "memory wall", the solution of Storage Class Memory (SCM) was first proposed by Intel Corporation. SCM acts as the bridge between DRAM and NAND, which has a higher storage density than DRAM and a higher access speed than NAND FLASH (Figure 1.2). To achieve high-density SCM, memories are often required to have the advantage of three-dimensional (3D) integration. In addition to the 3D stacking with conventional Flash memory cells, some emerging memories, such as PCRAM and RRAM, have intrinsic advantages for SCM applications.

According to the different integration methods, 3D crossbar arrays can be classified into two types. One is a 3D X-point structure, which is a multi-layer stack of planar crossbar arrays, and the other one is a vertical crossbar array structure (3D VRRAM), which is similar to BICS 3D NAND. Each layer of the crossed-array structure of the 3D X-point structure needs to be manufactured separately. The advantage of the 3D X-point structure is that the storage density per unit area can be improved while the 3D stacked structure is easy to integrate. For the vertical 3D structure of 3D VRRAM, the number of lithograph does not increase significantly with the number of stacking

DOI: 10.1201/9781003391586-1

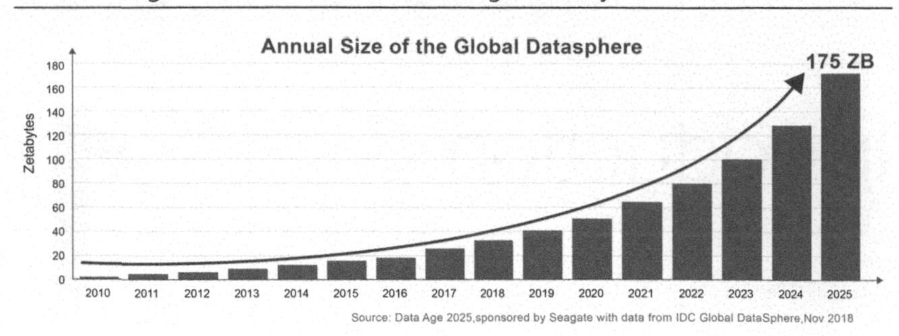

FIGURE 1.1 The total amount of global data.

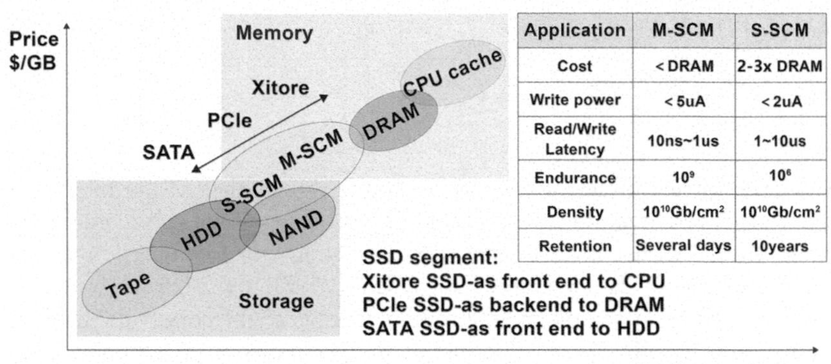

Application	M-SCM	S-SCM
Cost	< DRAM	2-3x DRAM
Write power	< 5uA	< 2uA
Read/Write Latency	10ns~1us	1~10us
Endurance	10^9	10^6
Density	10^{10}Gb/cm^2	10^{10}Gb/cm^2
Retention	Several days	10years

FIGURE 1.2 Storage class memory.

layers increased, which significantly reduces the number of lithographs. Compared to other multi-layer stacked structures, the costs can be significantly diminished for 3D VRRAM.

In crossbar array structures, due to the exit of leakage currents, the memory cell is usually connected in series with a selector to suppress the disturbing effects. For 3D X-point structures, the integration of the selector with the memory cell can be achieved easily by a planar process, whereas for 3D VRRAM structures, the integration of the selector is very difficult because in vertical arrays, the bottom electrode of each memory cell column is formed by a trench filling process and the selector cannot be integrated separately. Therefore, the use of highly non-linear self-passed memory cells is the key to the realisation of 3D VRRAM.

This book introduces the 3D integration technology of RRAM, and it basically breaks down into five parts:

Chapter 2: Crosstalk in Crossbar Array and 3D Architectures;
Chapter 3: Selector Devices and Self-Selective Cells;
Chapter 4: Integration of 3D RRAM;
Chapter 5: Reliability Issues of the 3D Vertical RRAM;
Chapter 6: Applications of 3D RRAM beyond Storage.

Crosstalk in Crossbar Array and 3D Architectures

Qing Luo

Contents

The crossbar array structure,[1–8] proposed as a memory architecture as early as six decades ago, is a structure composed of a series of horizontally parallel electrodes and some columns of longitudinal electrodes, with a layer of materials (RRAM, PCM, MRAM, etc.) as shown in Figure 2.1. Assuming that the line width of the upper and lower electrodes and the spacing between the parallel electrodes are both F, the area of each point, therefore, is $4F^2$. If such a structure is stacked in three dimensions, provided the number of stacked layers is L, the area of a single device is only $4F^2/L$. The crossed-array structure is an "ideal" option for the high-density integration of memories. Nevertheless, it is not widely adopted commercially. One important reason for this is that the applications of cross-array structures require device units that can meet many stringent requirements. At each connection point, the memory device needs to be able to store data while meeting the requirement of "nonlinearity", so that the state of unselected devices is not affected when the data is externally erased, and the data is not misread. The device units in a cross point-array structure must be two-terminal devices, with reliable, repeatable, and high on/off ratios.

DOI: 10.1201/9781003391586-2

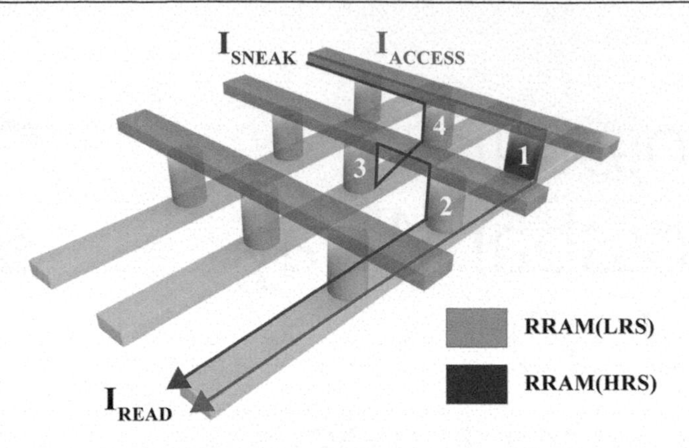

FIGURE 2.1 Crosstalk in the crossed array.

Over the last 10 to 15 years, memories with the aforementioned features have almost materialised. However, most two-terminal devices cannot meet the requirement of "nonlinearity". The crosstalk issues on failure to satisfy the nonlinearity requirement will be discussed in the following. Figure 2.1 illustrates a typical two-dimensional crossbar array. We intend to read the resistance state of device 1 that we selected, with its top and bottom electrodes applied with positive voltage and grounded, respectively. Assuming that device 1 is in a high resistance state (HRS) and devices 2, 3, and 4 are in a low resistance state (LRS), the red line marks a complete current loop, so the current we read is the current flowing through devices 2, 3, and 4 and the resistance state is read as a low resistance state. This misreading problem is known as crosstalk or sneaking current. Assuming there is an array with m rows and n columns, there will be $(m - 1) \times (n - 1)$ crosstalk loops, which means that the larger the array is, the more severe the crosstalk problem gets. In the worst case,[9] all unselected devices are in the low resistance state and only selected devices are in the HRS, making the problem much worse.

The reading crosstalk problem in large-scale crossbar arrays can be understood by analysing the worst-case array. As in Figure 2.2, assuming that there exists a crossed array with m rows and n columns and that the ratio of high and low resistance states of resistive devices is β, for resistors to be read as in an HRS, we assume that all other devices are in an LRS, and for resistors to be read as in an LRS, we suppose that all other devices are in an HRS. To avoid misreading, the HRS that we read must be higher than the LRS, i.e.,

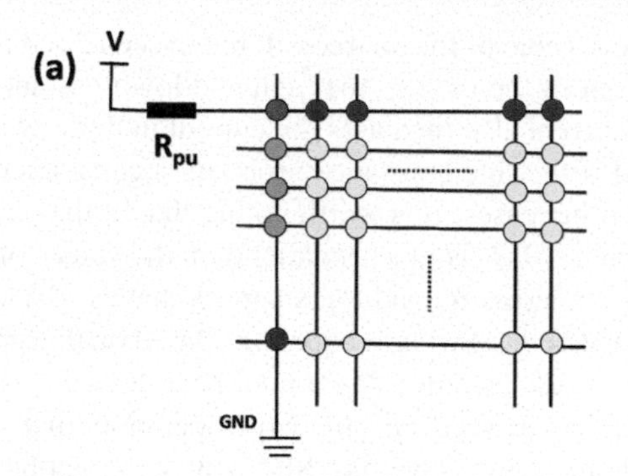

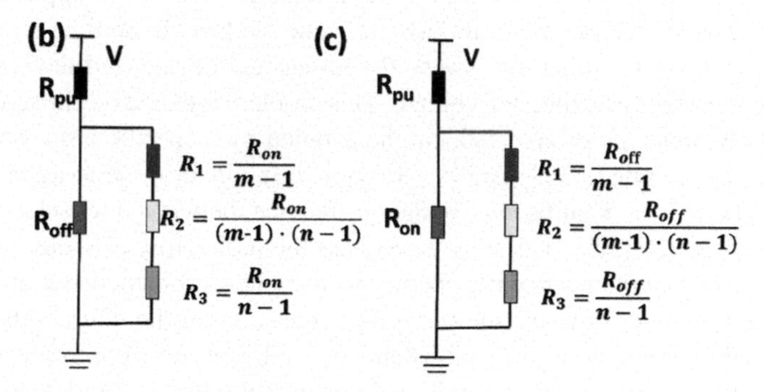

FIGURE 2.2 $m \times n$ Crossbar array. (a) Worst-case equivalent circuit. (b) (c) R_1, R_2, and R_3 are equivalent resistors for the selected row, unselected row, and selected column, respectively.

$$R_{\text{off}}//\left(\left(\frac{R_{\text{on}}}{m-1}\right)+\left(\frac{R_{\text{on}}}{(m-1)\cdot(n-1)}\right)+\frac{R_{\text{on}}}{n-1}\right) > R_{\text{on}}//\left(\left(\frac{R_{\text{off}}}{m-1}\right)+\left(\frac{R_{\text{off}}}{(m-1)\cdot(n-1)}\right)+\frac{R_{\text{off}}}{n-1}\right) \tag{2.1}$$

Hence

$$\left(1-\frac{m-1\cdot(n-1)}{m+n-1}\right)\cdot(\beta-1)>0 \tag{2.2}$$

Clearly the ratio of high and low resistance states β is greater than 1, so

$$n<\frac{2\cdot(m-1)}{m-2} \text{ or } m<\frac{2\cdot(n-1)}{n-2} \tag{2.3}$$

In other words, the maximum value of the matrix will not exceed 3×3 if n or m is greater than 2, independent of the ratio of the high and low resistance states β.[10] In addition, the sneak current also increases the current in the word and bit lines, which is exacerbated if the line resistance problem is considered,[11] and the power consumption also increases. It is worth noting that in the equivalent circuit in Figure 2.2b, the value of R_2 is much smaller than the values of R_1 and R_3, and the voltage allocation across R_1 and R_3 is approximately $V/2$ (V is the voltage applied to the selected device) when n equals m. Based on this principle, it is possible to control the sneak current using a nonlinear device.

The same crosstalk problem exists during the write operation. In the case of a 2×2 crossbar array (Figure 2.3a), when the SET voltage is applied to the selected device (red device), almost the same voltage is applied to several unselected devices, which may cause the unselected device (red device) to be operated incorrectly by the SET voltage. This problem is called writing crosstalk. Fortunately, this type of crosstalk can be avoided by a specific write operation mode, e.g. $V/2$ write or $V/3$ write[12,13] (Figure 2.3). Under these write operation modes, there is a significant voltage difference between the selected and unselected devices, which effectively reduces the probability of write crosstalk. However, the two modes of write operation raise a new problem on reliability for unselected memory devices: the write disturbance issue, i.e. the ability of the unselected region to constantly withstand 1/2 or 1/3 of the write voltage.

To eliminate the read crosstalk and enhance the integrated density of the memory, one ideal way is to connect a two-terminal selector to the two-terminal memories in a series. The functions of the selector are to reduce the leakage current in the potential sneak paths. This function is realised in light of two important features of the sneak paths. First, at least one device in the conductive path is inversely conductive. Second, the full voltage is applied to at least three devices in series.

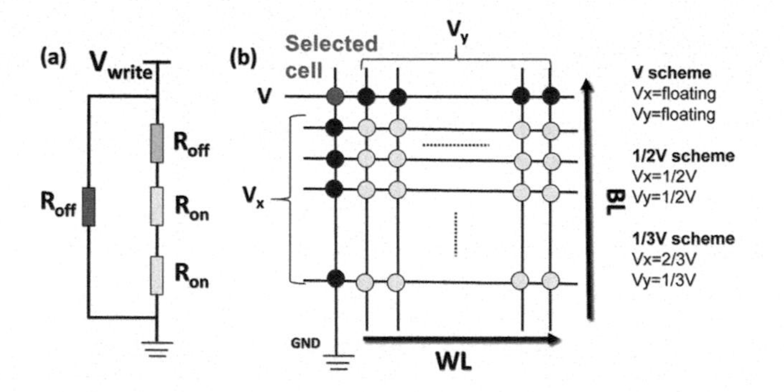

FIGURE 2.3 (a) Equivalent circuit scheme of the 2 × 2 crossbar array. (b) Scheme of the write operation voltage configuration.

This means that there are at least two sorts of selectors available that can get rid of the read crosstalk. One type of selector can allow the forward voltage to pass through to suppress the reverse voltage, which are asymmetrical devices, such as diodes; the other type of selector is turned on at high voltage and closed at low voltage, which is a nonlinear device, such as mixed-ionic electron conduction (MIEC) device. Besides, automotive rectifier devices, self-gated devices, and complementary resistive switching devices can suppress leakage currents as well.

To support high-density integration of memory crossbar arrays, an ideal selector should have a range of necessary features. And of them, some major ones are shown in the following article.

1. *On-state current density*: To conduct the current for the memory to erase and write, each selector connected in series on the memory must be able to withstand high currents (no breakdown). To conduct a high current over a small cross-sectional area, the device needs to be able to bear a high current density (>1 MA/cm^2).
2. *Off-state leakage current*: The leakage current in unselected devices needs to be remarkably smaller than the on-state current, to effectively control the leakage current through all unselected devices.
3. *Nonlinearity*: On/off current ratio. Nonlinearity indicates the ability of the selector to limit leakage current. The larger the nonlinearity is, the larger the size of the crossbar array can be. Nonlinearity is defined as $I@V_{read}/I@1/2V_{read}$ (assuming the read voltage is V_{read}, $I@V_{read}$ indicates the current value at voltage V_{read}.)
4. *Bidirectional operation*: RRAM devices with the bidirectional operation have better fatigue resistance and lower power consumption. As the selector is used in a crossbar array that can be applied to various types of memory, it needs to support bidirectional operation.
5. *Back end of line (BEOL) compatibility*: BEOL compatibility means that the device must be made under less than 400 °C during the manufacturing process, and the selector must be able to withstand the high temperature of more than 400 °C for more than two hours.
6. *Others*: Because the characterisation of the selector must not limit the characterisation and reliability of the storage unit, all other properties of the selector must outperform those of the memory cells. These properties include turn-on speed, fatigue characteristics, yield in the array, uniformity, etc.

The above requirements explain why it is such a demanding task to produce a perfectly suitable selector that can be applied to large-scale 3D integration of resistance-type memory. Next chapter will introduce a series of reported selectors and analyse their advantages and disadvantages.

REFERENCES, BIBLIOGRAPHY, OR WORKS CITED

[1] G.W. Burr, R.S. Shenoy, and H. Hwang, "Select device concepts for crossbar arrays," *Resistive Switching: From Fundamentals of Nanoionic Redox Processes to Memristive Device Applications*, pp. 623–660, 2016.

[2] W.E.P. Goodwin, Electric connecting device: U.S. Patent 2,667,542[P]. 1954-1-26.

[3] M.M. Ziegler, and M.R. Stan, "Design and analysis of crossbar circuits for molecular nanoelectronics," *Proceedings of the 2nd IEEE Conference on Nanotechnology*, Washington, DC, USA, 2002, pp. 323–327.

[4] I.G. Baek et al., "Multi-layer cross-point binary oxide resistive memory (OxRRAM) for post-NAND storage application," *IEEE International Electron Devices Meeting, 2005. IEDM Technical Digest.*, Washington, DC, USA, 2005, pp. 750–753.

[5] M.-J. Lee et al., "2-stack 1D-1R Cross-point Structure with Oxide Diodes as Switch Elements for High Density Resistance RAM Applications," *2007 IEEE International Electron Devices Meeting*, Washington, DC, USA, 2007, pp. 771–774.

[6] Y. Bai, H. Wu, R. Wu, et al., "Study of multi-level characteristics for 3D vertical resistive switching memory[J]," *Scientific Reports*, vol. 4, p. 5780, 2014.

[7] S. Yu, H.Y. Chen, B. Gao, et al., "HfO_x-based vertical resistive switching random access memory suitable for bit-cost-effective three-dimensional cross-point architecture[J]," *ACS Nano*, vol. 7, no. 3, pp. 2320–2325, 2013.

[8] W.C. Chien et al., "Multi-layer sidewall WOX resistive memory suitable for 3D ReRAM," *2012 Symposium on VLSI Technology (VLSIT)*, Honolulu, HI, USA, 2012, pp. 153–154.

[9] J.Y. Seok, S.J. Song, J.H. Yoon, et al., "A review of three-dimensional resistive switching cross-bar array memories from the integration and materials property points of view[J]," *Advanced Functional Materials*, vol. 24, no. 34, pp. 5316–5339, 2014.

[10] A. Flocke and T.G. Noll, "Fundamental analysis of resistive nano-crossbars for the use in hybrid Nano/CMOS-memory," *ESSCIRC 2007 - 33rd European Solid-State Circuits Conference*, Munich, Germany, 2007, pp. 328–331.

[11] C.L. Lo, T.H. Hou, M.C. Chen, et al., "Dependence of read margin on pull-up schemes in high-density one selector–one resistor crossbar array[J]," *IEEE Transactions on Electron Devices*, vol. 60, no. 1, pp. 420–426, 2013.

[12] S. Yu, J. Liang, Y. Wu, et al., "Read/write schemes analysis for novel complementary resistive switches in passive crossbar memory arrays[J]," *Nanotechnology*, vol. 21, no. 46, p. 465202, 2010.

[13] P.O. Vontobel, W. Robinett, P.J. Kuekes, et al., "Writing to and reading from a nano-scale crossbar memory based on memristors[J]," *Nanotechnology*, vol. 20, no. 42, p. 425204, 2009.

Selector Devices and Self-Selective Cells

3

Yaxin Ding and Qing Luo

Contents

DOI: 10.1201/9781003391586-3

3.1 NONLINEAR SELECTOR DEVICES

Several access devices such as transistors and diodes have been proposed to solve the leakage current and crosstalk problems of the memory array with a crossbar array.[1–5] However, transistor-based access devices have a large cell size, which can hinder the high-density integration of memory. The stackable one diode–one resistor (1D1R) cell structure is considered one of the most attractive solutions because it is compatible with the $4F^2$ crossbar array.[6,7] But the diode-based access devices, which are not unsuited to bipolar memories, can only be operated in one direction. Only selector devices with both high nonlinearity and fast switching speed can be a potential candidate for constructing three-dimension memory arrays. Moreover, ideal selector devices have several other critical characteristics such as low off-state leakage current, high on-state current density, bidirectional switching, and so on.[8–10] The low off-state current can suppress the leakage current and crosstalk effect of the memory array. The high on-state current density can supply high drive current to programme and erase the memory element. The selector devices need to be capable of bidirectional switching because most memories need to operate in bipolar mode. So far, various selector devices have been widely studied for high-density memory arrays.

3.1.1 Tunnel Barrier-Based/Type Selector

Uniformity is critical for the selector due to the read region of the selector is highly dependent on the threshold switching voltage range (V_{th} min to V_{th} max).[11] In order to read the memory array successfully under the commonly used $V/2$ bias configuration, the minimum read voltage window should be smaller than the V_{th} min.[12] When pure electron conduction and the absence of ion movement or structure modification occur during the operation, tunnel barrier stacks show terrific uniformity.[13,14] Thus, the tunnel barrier-based selective device with excellent uniformity can be a promising candidate for the selector, which can be created by using a thin oxide layer (HfO_2, TaO_x, ZrO_2, and TiO_2)[15,16] or a nitride layer as a tunnelling barrier. The tunnel barrier can be engineered by stacking multilayer dielectrics. The design of the multilayer dielectrics is critical for creating a selection device with both high uniformity and high nonlinearity because the nonlinearity of the interface type selector is not as satisfying as that of a filamentary type device. Huang et al.[13] fabricated a bipolar nonlinear selector with a rectangular energy band structure by using a simple Ni/TiO_2/Ni metal–insulator–metal stacking (Figure 3.1a). As shown in Figure 3.1b, over six

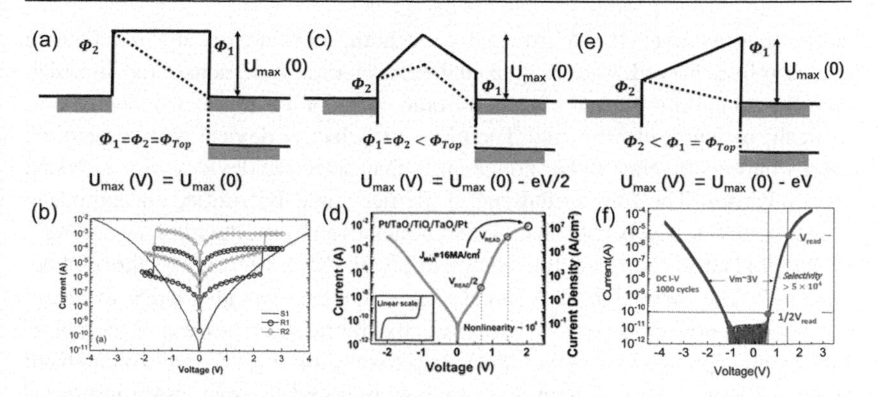

FIGURE 3.1 Schematic diagram of (a) rectangular energy band structure and (b) IV characteristics of the bipolar Ni/TiO_2/Ni MIM selector; (c) crested band structure and (d) IV characteristics of a Pt/TaO_x/TiO_2/TaO_x/Pt stack; (e) trapezoidal band structure and (f) typical I-V characteristics of the TaO_x-based selector. 3.1(a, c, e) Reprinted/adapted by permission from [Springer Nature Customer Service Centre GmbH]: [Nano Research, Springer Nature] [Highly uniform and nonlinear selection device based on trapezoidal band structure for high density nano-crossbar memory array] by [Qing Luo et al.][COPYRIGHT] (2017). 3.1(b) © [2011] IEEE. Reprinted, with permission, from [Jiun-Jia Huang, Bipolar Nonlinear Ni/TiO_2/Ni Selector for 1S1R crossbar array applications, IEEE Electron Device Letters, and Oct. 1, 2011]. 3.1(f) © [2016] IEEE. Reprinted, with permission, from [Qing Luo, Fully BEOL compatible TaO_x-based selector with high uniformity and robust performance, IEEE Proceedings, and Dec. 1, 2016]. 3.1(d) Reprinted (adapted) with permission from [W. Lee, et al., "High current density and nonlinearity combination of selection device based on TaO_x/TiO_2/TaO_x structure for one selector–one resistor arrays," ACS Nano, vol. 6, no. 9, pp. 8166–8172, Sep. 25, 2012]. © [2012] American Chemical Society.

orders of magnitude of current increase for a voltage swing from 0 to ±2 V are realised and breakdown voltage larger than 4 V is also achieved. But the maximum barrier height (U_{max}) changes slowly in the rectangular energy band structure with the uniform barrier, which is unable to meet the requirements of highly nonlinear selector devices. Lee et al.[14] demonstrated a high-performance selection device based on TaO_x/TiO_2/TaO_x structure. As shown in Figure 3.1c, the energy band of the TiO_2 film is symmetrically bent at the top and bottom TaO_x/TiO_2 interfaces due to the gradual diffusion of some Ta atoms into the TiO_2 film, thus achieving a crested oxide barrier. Compared with the rectangular energy band structure, the highest part of the crested energy band structure (in the middle) is pulled down by the electric field more quickly ($U_{max}(V) = U_{max}(0) - eV/2$).[12] As a result, the tunnelling current through a crested energy barrier could increase dramatically compared with the tunnelling current through a uniform barrier, which remarkably improves nonlinearity. As shown in Figure 3.1d, high current density over 10^7 A/cm^2 and outstanding nonlinearity up to 10^4 were successfully achieved. And a trapezoidal energy barrier is demonstrated, in which both the tunnelling current and thermionic emission current can increase dramatically under the electric field. Therefore, the selective devices with trapezoidal energy barriers display higher nonlinearity than selective devices with a crested energy barrier. The trapezoidal energy barrier could be formed in composite semiconductors. The barrier shape could be achieved by modulating the doping[17] or by gradually changing the composition of the layer during the epitaxy process.[18] The formation of "staircase" potential patterns also helps to obtain trapezoidal energy barriers.[16] However, the fabrication process is complex. Luo et al.[11] proposed a novel TaO_x-based selector with a trapezoidal band structure (Figure 3.1e), which was obtained by rapid thermal annealing in O_2 plasma. Ta is fully oxidative on the surface of the TaO_x layer. As the depth increases, the component of oxygen and the band gap of the TaO_x film decrease. And eventually, a trapezoidal band structure is achieved. The device shows robust performance such as high current density (1 MA/cm^2), high selectivity (5×10^4), low off-state current (~10 pA), robust endurance (>10^{10}), self-compliance and excellent uniformity (Figure 3.1f), suggesting that the tunnel barrier-based selective device could turn high-density 3D RRAM storage into a reality.

3.1.2 Metal Conductive Filament-Based Threshold Switching Selector

Selective devices with low leakage currents are essential for suppressing crosstalk suppression and lowering power consumption.[19] Therefore, metal conductive filament-based threshold switching selectors have made their presence known because of their ultra-low leakage current and high nonlinearity. Ag is one of the most common active electrode materials for constructing threshold switching (TS) selective devices owing to its outstanding oxygen solubility and diffusivity.[20] Song et al.[21] proposed an Ag/TiO_2-based threshold selector device with excellent selector characteristics such as high selectivity (~10^7), low leakage current (<10 pA), and steep slope (<5 mV/decade). Typical threshold switching behaviour is shown in Figure 3.2a. And the device suddenly switches to the low resistance state (LRS) when the applied voltage exceeds the threshold switching voltage (V_{th}). Then the device switches back to the off state when the applied voltage is smaller than the hold voltage (V_{hold}). As shown in Figure 3.2b, the spontaneous rupturing of silver (Ag) filament is responsible for the volatile threshold switching (TS) behaviour.

The size and shape of the filament are vital to the switching characteristic of Ag-based filament devices. The thin filament will incur the volatile TS

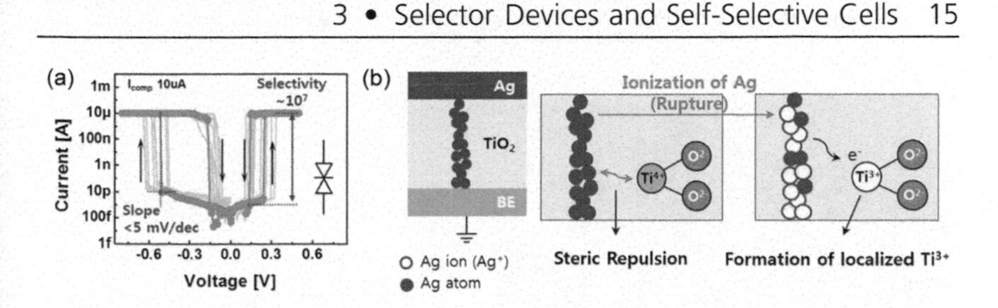

FIGURE 3.2 (a) Threshold switching behaviours of Ag/TiO_2-based selector device. (b) The switching mechanism of threshold switching behaviour in Ag/TiO_2-based selector device.

behaviour and the thick filament will incur the non-volatile memory switching (MS) behaviour. What's more, the excessive diffusion of Ag ions usually causes failure of stuck to on-state, which often leads to switching reliability issues.[22] The diffusion of Ag in the film can be controlled by inserting a graphene barrier or metal barrier.[22] In addition, a previous study proposed that the amount of Ag inside the structure can be reduced by engineering Ag nanodots.[23] And yet the precise control of Ag diffusion is still under study. Banerjee et al.[20] demonstrated a method to control precisely the Ag diffusion in Ag/HfO_x/Pt devices through a vacancy-induced-percolation (VIP) path. The high-resolution transmission electron microscopy (HRTEM) image is shown in Figure 3.3a. The Ag ions can diffuse through the localised path caused by the nonstoichiometric HfO_x film with the pre-existing vacancy (Figure 3.3c). Highly stable TS behaviour such as low fluctuation (<3%), low slope (<2 mV/dec), ultra-low off-current (~0.4 pA), high selectivity ($>4 \times 10^{10}$), and device yield of 100% is achieved in the Ag-filament based selector (Figure 3.3b). Furthermore, the device can achieve an endurance of 10^9 at a selectivity of 10^8 (Figure 3.3d).

However, the Ag-based selective devices still suffer from the CMOS incompatibility issue despite their outstanding advantages. In contrast, Cu can be a better choice as another common active electrode material. Luo et al.[19] demonstrated a fully CMOS-friendly Cu-doped HfO_2 material-based selector with low leakage current and high nonlinearity. As shown in Figure 3.4a, a Cu-doped HfO_2 film is used as the threshold switching layer and a thin HfO_2 film serves as the tunnelling layer to reduce the leakage current. Compared with the device with a single TS layer, the device with a tunnelling layer exhibits more than five orders of magnitude decrease in the leakage current (Figure 3.4b). The observed threshold switching behaviour is caused by the spontaneous rupture of conductive filament in doped HfO_2 film, which is the same as that of the Ag-based selective devices. In addition, the selector device shows robust performance such as high nonlinearity (~10^7), ultra-low

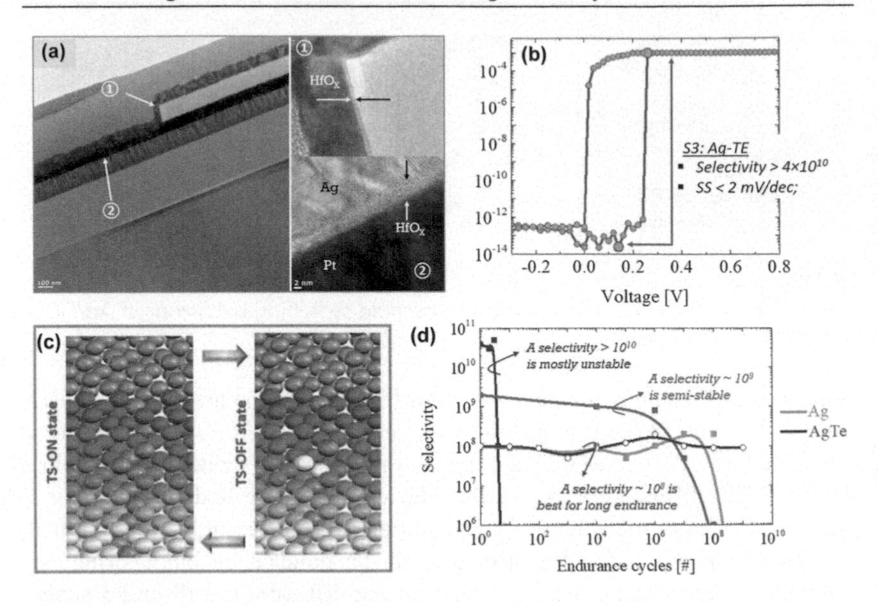

FIGURE 3.3 (a) High-resolution transmission electron microscopy image of Ag/HfO_x/Pt device. (b) IV characteristics of the selector shows excellent threshold switching performance. (c) The switching mechanism in the TS mode. (d) The selector can achieve an endurance of 10^9 at a selectivity of 10^8. © [2021] John Wiley and Sons. Reprinted, with permission, from [Hyunsang Hwang, Donghwa Lee, Sangmin Lee, et al., Deep Insight into Steep-Slope Threshold Switching with Record Selectivity ($>4 \times 10^{10}$) Controlled by Metal-Ion Movement through Vacancy-Induced-Percolation Path: Quantum-Level Control of Hybrid-Filament, Advanced Functional Materials, and Jun. 26, 2021].

leakage current (~pA), high endurance ($>10^{10}$), and sufficient on-state current density (~1 MA/cm^2) (Figure 3.4c, d).

3.1.3 Metal Insulator Transition Selector

The metal-insulator-transition (MIT) behaviour occurring in transition metal oxides such as niobium oxide (NbO_x) and vanadium oxide (VO_x) has been extensively studied. The MIT device will change from an insulating state to a metallic state after applying certain external stimuli such as temperature,[24] optical,[25] or electric field.[26] Electrically driven MIT devices are well-suited selectors due to their fast transition speed and bidirectional switching characteristics. The threshold switching characteristic of the MIT selector is like that of the metal conductive filament-based threshold switching selector. The device changes from the insulating state to the metallic state when the applied

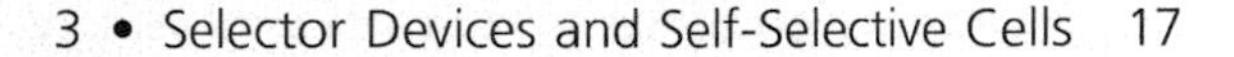

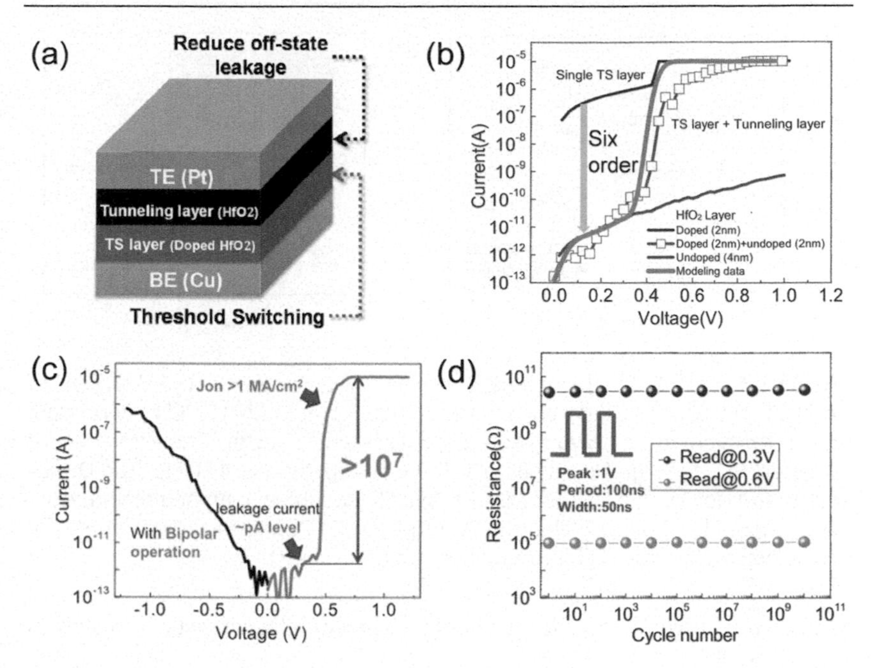

FIGURE 3.4 (a) Bilayer device structure by introducing a tunnelling layer on the TS layer. (b) I-V curves of devices with single TS layer, bilayer, and undoped layer. (c) I-V characteristic of the bilayer selector device with the optimised tunnelling layer thickness. (d) Endurance test of the bilayer selector device. © [2015] IEEE. Reprinted, with permission, from [Qing Luo, Cu BEOL compatible selector with high selectivity ($>10^7$), extremely low off-current (~pA) and high endurance ($>10^{10}$), IEEE Proceedings, and Dec. 1, 2015].

voltage on the device surpasses the V_{th}. And it turns back to the insulating state when the applied voltage on the device is smaller than V_{hold}.

Vanadium dioxide (VO_2) suffers from first-order insulator-metal transition due to the joule heating after applying an electric field. The VO_2 changes from the insulating state to the metallic state when the temperature rises because joule-heating surpasses the transition temperature (TC = 340 K = 67 °C). Son et al.[25] proposed a nanoscale VO_2 device with high a on/off ratio (>50) and high current density ($>10^6$ A/cm^2). The I-V characteristics of the VO_2 device are shown in Figure 3.5a. The inset on the right shows the structure of the nanoscale Pt/VO_2/Pt device. In addition, the nanoscale VO_2 devices show a fast response (<20 ns) to an input voltage signal and stable hysteretic I-V characteristics. However, the low transition temperature of VO_2 hampers its application because standard operating temperatures exceed 90 °C[27] Rupp et al.[27] showed the I-V characteristics of the Pt/VO_2/Pt device cycled at different temperatures.

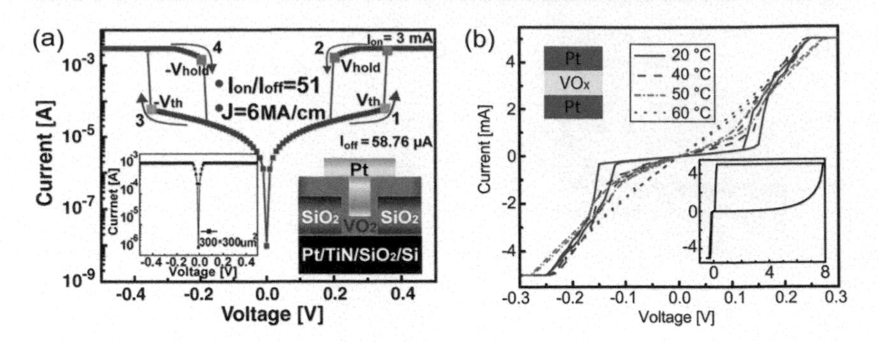

FIGURE 3.5 (a) The I-V characteristics of the Pt/VO_2/Pt device. (b) I-V sweeps of the Pt/VO_2/Pt device at different temperatures. 3.5(a) © [2011] IEEE. Reprinted, with permission, from [Myungwoo Son, Excellent Selector Characteristics of Nanoscale VO_2 for High-Density Bipolar ReRAM Applications, IEEE Electron Device Letters, and Nov. 1, 2011]. 3.5(b) © [2016] IEEE. Reprinted, with permission, from [J.A.J. Rupp, Threshold Switching in Amorphous Cr-Doped Vanadium Oxide for New Crossbar Selector, IEEE Proceedings, and May 1, 2016].

As shown in Figure 3.5b, the threshold switching characteristic vanishes at roughly 60 °C.

NbO_2 has a high bulk transition temperature (TC = 1,070 K = 797 °C) compared with VO_2. A nanoscale device with ultrathin NbO_2 film fabricated by the reactive sputtering method has been studied for selective device applications.[28,29] Cha et al.[30] demonstrated a NbO_2 selector with reduced leakage current by adopting a 10-nm-thick TiN bottom electrode with low thermal conductivity. Figure 3.6a shows the 3D device structure of the W/NbO_2/TiN device. The leakage current of the 3D vertical device is significantly reduced compared with the planar device due to the heat confinement effect (Figure 3.6b). NbO_2 MIT-based selective devices exhibit well-behaved hysteretic I-V characteristics like VO_2 devices. The temperature dependence of NbO_2 devices before and after threshold switching at different read voltages is shown in Figure 3.6c. Stable threshold switching characteristics are still observed at 125 °C, sufficient for reliable memory application. However, the endurance of the NbO_2 selector with 3D vertical architecture is only 10^6. And the endurance of the selector should be higher than 10^9 as the write endurance of memory type SCM is high ($>10^9$). In addition, the leakage current of the MIT selector is relatively high compared with other selector candidates such as metal conductive filament-based threshold switching selector and ovonic threshold switching (OTS) selector. Defects exist in MIT material that generate the conduction subbands between the conduction band and valance band, which leads to the high leakage current of the MIT selector. In addition, the interface defects between an electrode and IMT materials can

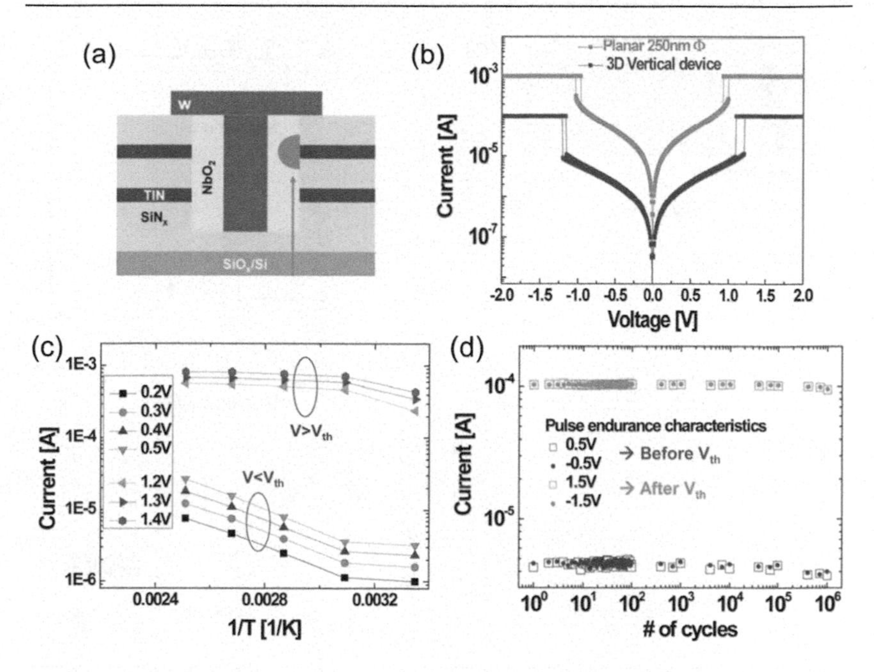

FIGURE 3.6 (a) The structure of the W/NbO_2/TiN device. (b) The leakage current is reduced by adopting a 3D structure. (c) Thermal stability test of the NbO_2 threshold device before and after threshold switching. (d) The endurance test of the NbO_2 selector. © [2013] IEEE. Reprinted, with permission, from [Euijun Cha, Nanoscale (~10 nm) 3D vertical ReRAM and NbO_2 threshold selector with TiN electrode, IEEE Proceedings, and Dec. 1, 2013].

further increase the leakage current because the defects can pin the Schottky barrier height.[31] Inserting a barrier layer between the electrode and IMT film can eliminate both interfacial and bulk defects.[31] Furthermore, the leakage current can be decreased by adding a barrier layer.

Luo et al.[32] proposed an $Nb_{1-x}O_2$-based selector with ultra-high endurance ($>10^{12}$), high operation speed (10 ns), bidirectional operation, and excellent V_{th} stability. As shown in Figure 3.7a, the TEM image of the selector shows its 3D structure. The XPS fitting result of the NbO_x film is shown in Figure 3.7b. As shown in Figure 3.7c, the I-V characteristic of the selector shows 60x non-linearity and a high on-state current density (4.8 MA/cm^2). Stable threshold switching behaviour is observed during the 10^{12} cycles endurance test (Figure 3.7d). The off-state leakage current is reduced by one order of magnitude (selectivity as high as 500) by adding a barrier layer between $Nb_{1-x}O_2$ film and electrode (Figure 3.7e). Nevertheless, the NbO_x based selector device still has some drawbacks such as high leakage current and low nonlinearity.

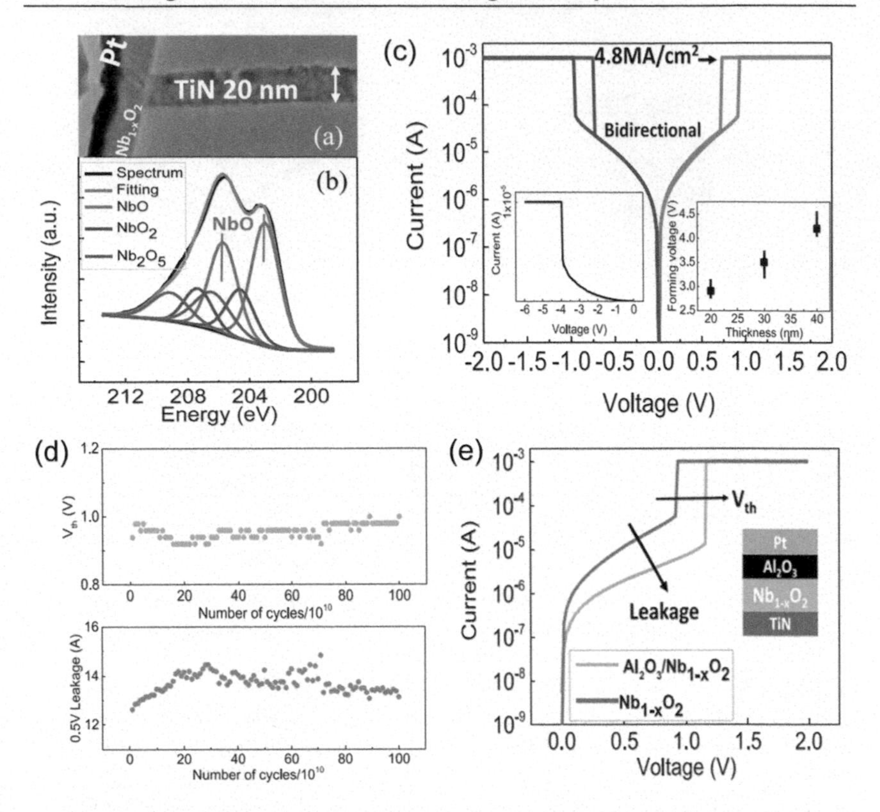

FIGURE 3.7 (a) TEM image of the selector. (b) Threshold switching characteristic of the device. (c) The endurance test of NbO_x threshold device. (d) The leakage current is decreased by adding a barrier layer (Al_2O_3). (e) I-V curves of devices with single TS layer and $Al_2O_3/Nb_{1-x}O_2$ bi-layer design. Leakage is reduced by barrier layer.

3.1.4 Ovonic Threshold Switch Selector

The ovonic threshold switching (OTS) phenomenon was first reported in thin films of amorphous chalcogenide alloys by S.R. Ovshinsky.[33] So far, nano-scale OTS device with a T-shaped structure has been widely investigated.[34,35] Compared with the various TS type selectors, the most favourable factor of the OTS selector is that it can satisfy the requirements of most selective devices thanks to its comprehensively superior performance. The off-state resistance of the OTS selector is much higher than that of the MIT selector. And the on-state current of the OTS selector is sufficient to program and erase the memory element. In addition, ps-level transition speed is achieved in the OTS selector.[36] Previous studies[37,38] have demonstrated that OTS selector

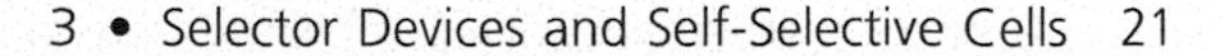

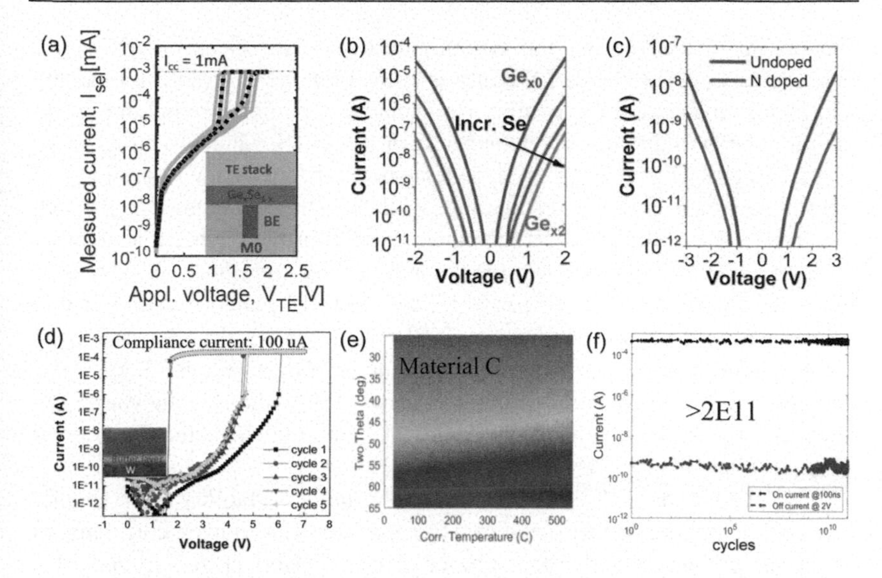

FIGURE 3.8 (a) DC threshold switching characteristic of the GeSe device; The leakage current is decreased by (b) increasing Se content (c) or doping element N; (d) DC I-V curves of the AsSeGeSi device shows ultra-low leakage current; (e) Thermal stability is verified by XRD; (f) The device is alive even after 2E11 cycles. 3.8(a, b) © [2018] IEEE. Reprinted, with permission, from [Naga Sruti Avasarala, Half-threshold bias I_{off} reduction down to nA range of thermally and electrically stable high performance integrated OTS selector, obtained by Se enrichment and N-doping of thin GeSe layers, IEEE Proceedings, and Jun. 1, 2018]. 3.8(c–f) © [2020] IEEE. Reprinted, with permission, from [H.Y. Cheng, Si Incorporation into AsSeGe Chalcogenides for High Thermal Stability, High Endurance and Extremely Low V_{th} Drift 3D Stackable Cross-Point Memory, IEEE Proceedings, and Jun. 1, 2020].

exhibited electronic type transmission with possible secondary thermal effects, suggesting that it does not involve any atomic arrangements while switching, which can explain why OTS selector perform so well, such as ultra-fast switching speed and superb switching endurance. Therefore, the OTS selector is very suitable for selector device applications.

OTS selectors can be classified into two categories: the Se-based OTS selector and the Te-based OTS selector. The Se-based OTS selector shows lower leakage current than the Te-based device.[36] Govoreanu et al.[35] demonstrated a Se-based OTS selector with a record drive current densities (exceeding 20 MA/cm^2) and high thermal reliability (350 °C). However, only 3,500 half-bias nonlinearity is shown in the device (Figure 3.8a). Avasarala et al.[39] showed that the leakage current at half threshold bias ($I_{off1/2}$) was decreased to the 1nA range by using Se-enriched or N-doped GeSe

(Figure 3.8b & c). And the half-bias nonlinearity of the $Ge_{x1}Se_{1-x1}N$ device increases to 10^5. The thermal stability is critical for the OTS selector as only amorphous chalcogenides exhibit the volatile threshold switching behaviour.[40] The introduction of elements such as As, Si, and N can improve thermal stability.[41,42] Cheng et al.[34] demonstrated that Si incorporation can improve thermal stability and endurance while also achieving good I_{off}, which effectively relaxes the trade-off that attempts to improve the thermal stability of AsSeGe systems and degrade the I_{off} and cycling endurance. As shown in Figure 3.8d, ultra-low I_{off} (18 pA@2V) is achieved by incorporating Si into the AsSeGe system. In addition, the XRD result shows that the AsSeGeSi selector maintains an amorphous state even up to 550 °C (Figure 3.8e). Pulse with on current (~300 μA) at 100 ns is used to test the AC switching endurance of the AsSeGeSi selector. Threshold switching characteristics are not degraded after 2E11 cycles (Figure 3.8f).

The Te-based OTS selector has better device stability and smaller threshold switching voltage (which is critical for low power consumption operation) compared with the Se-based device.[36] Koo et al.[43] proposed a simple binary SiTe OTS device with excellent performance such as high selectivity (~10^6) and fast operating speed (2 ns transition after 10 ns delay) (Figure 3.9b). However, the poor switching endurance characteristic (500k) is one of the main limitations of such Te-based binary system (Figure 3.9c).

Methods to improve endurance characteristics of the OTS selector have attracted tremendous interest. Garbin et al.[42] demonstrated a quaternary Si-Ge-As-Te OTS material system, which has been studied since the 1960s.[33] The crystallisation temperature increases (T_X > 450 °C) by adding Si to the Ge-As-Te system, which is related to the increase of the optical band gap

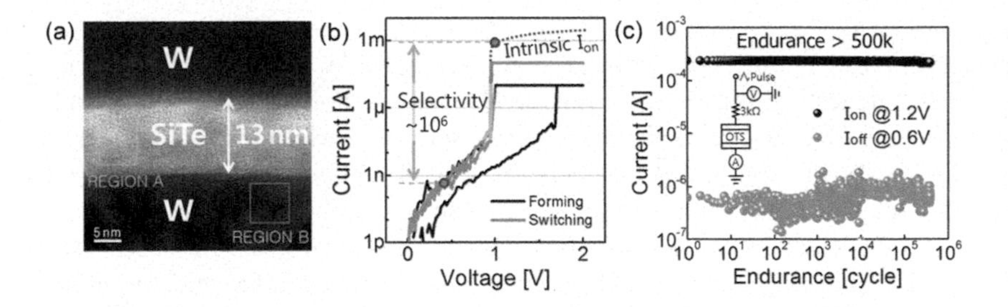

FIGURE 3.9 (a) TEM image of the SiTe binary OTS selector device; (b) DC I-V curves of the SiTe device shows ultra-high selectivity; (c) AC switching endurance of the device. © [2016] IEEE. Reprinted, with permission, from [Yunmo Koo, Te-based amorphous binary OTS device with excellent selector characteristics for x-point memory applications, IEEE Proceedings, and Jun. 1, 2016].

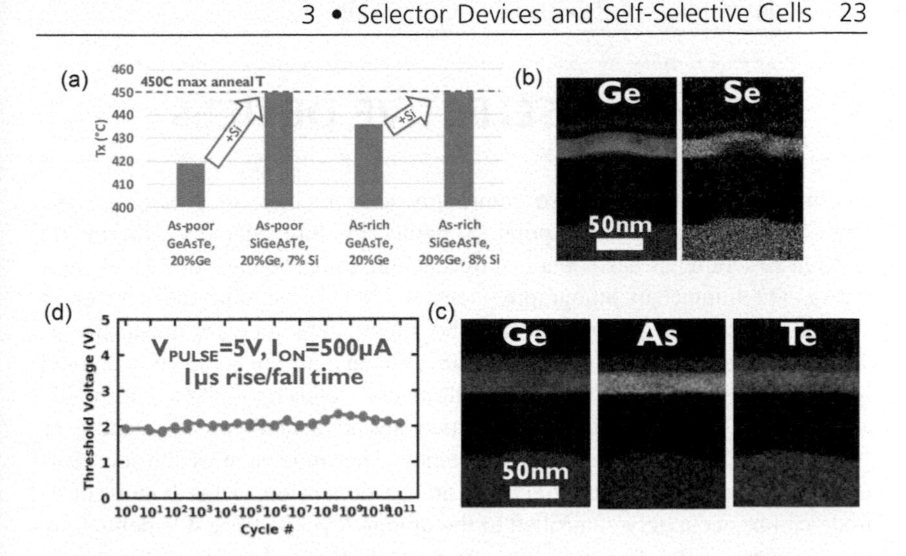

FIGURE 3.10 (a) In situ XRD analysis shows crystallisation temperature for different compositions. (b) Elemental mapping of GeSe OTS device after 1e8 cycles shows that the elemental segregation is responsible for the failure. (c) Elemental mapping of SiGeAsTe OTS device after 1e8 cycles shows uniform elemental distribution. (d) Stable endurance of 1e11 cycles is achieved in SiGeAsTe OTS device. © [2019] IEEE. Reprinted, with permission, from [D. Garbin, Composition Optimization and Device Understanding of Si-Ge-As-Te Ovonic Threshold Switch Selector with Excellent Endurance, IEEE Proceedings, and Dec. 1, 2019].

(Figure 3.10a). Furthermore, the failure mechanism is studied to improve the endurance characteristic of the OTS system. TEM analysis shows that the elemental segregation is responsible for the sudden failure of the GeSe device (Figure 3.10b). And the uniform elemental distribution is shown in the elemental mapping of the SiGeAsTe OTS device after 1e8 cycles, which is related to the formation of stable bonds (Figure 3.10c). Robust switching endurance of 1e11 cycles is achieved by incorporating Si into ternary Ge-As-Te device (Figure 3.10d).

The OTS selector has been a promising candidate for the commercialisation of selector devices due to its satisfactory characteristics. Elemental doping has been widely used in OTS selectors to enhance the device performance such as thermal reliability and cycling endurance. However, complex material compositions (four to five elements) caused by doping will induce compositional inhomogeneities.[40] Even worse, the reduced dimensions will degrade the thermal stability caused by interfacial hetero-crystallisation.[44,45] Therefore, the complex compositions caused by elemental doping hinder its application in selective devices.

3.2 SELF-SELECTIVE DEVICES

Nonlinear selector devices are commonly used in memories based on the three-dimensional (3D) cross-point (X-point) structure. Memory with the 3D X-point structure has been obtained by stacking the 2D cross-point for several layers. The number of lithography steps is $2N + 1$ assuming the number of stacked layers is N. Therefore, the lithography steps increase drastically as stacked layers rise, because the active area dimensions are completely defined by lithography in the 3D X-point based memory, causing high cost of the 3D X-point structure. In contrast, 3D vertical resistive random access memory (3D VRRAM) shows better characteristics. The dimension of the top electrode is determined by lithography. And the dimension of the bottom electrode can be accurately controlled to the atomic level because it is defined by the thickness of the deposited film. As a result, the uniformity of the device can be improved thanks to the precise fabrication process. In addition, the number of critical lithography steps does not increase as the number of stacked layers increases, suggesting lower cost and higher density.

The memory cells are one time deposited on the side wall of the deep holes in a 3D VRRAM architecture. As shown in Figure 3.11a, an intermediate electrode exists between the selective layer and the bit line (BL), which causes a short circuit in the memory cell on the same BL by connecting the same selector. Therefore, the 1S1R structure is not suitable to 3D VRRAM. Instead, the self-selective memory cell with rectifying or built-in nonlinearity characteristic (the current change nonlinearly with the voltage at LRS) is the only choice for constructing a 3D VRRAM. The construction of the self-selective memory cell is shown in Figure 3.11b, which commonly is comprised of a selective layer and a memory layer.

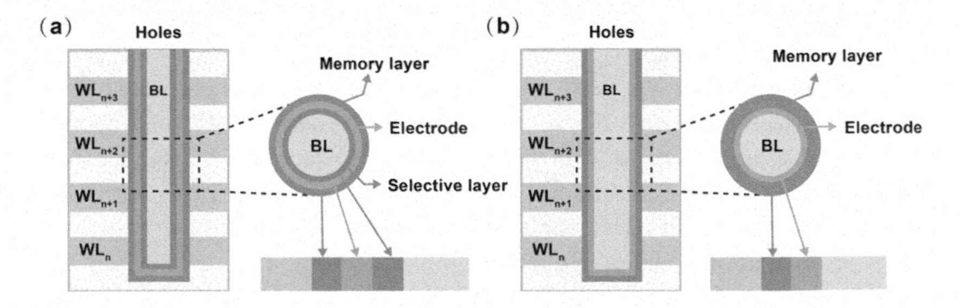

FIGURE 3.11 (a) The memory cells on the same BL will be shorted due to the introduction of an intermediate electrode. (b) The self-selective memory cell consists of a selective layer and a memory layer.

3.2.1 Self-Rectifying RRAM

The self-rectifying device can be used to inhibit crosstalk in a crossbar memory without extra rectifying diodes as the I-V characteristic curve of the self-rectifying device is like that of the 1D1R. Previous studies have reported that the memory with self-rectifying effect by using Al/PCMO/TiN structure[46] or HfO_x/ZrO_x-based RRAM with a back-to-back connection technique.[47] However, these memories have downsides such as low rectifying behaviour, high read-out voltage, and complex device structure. The formation of a Schottky barrier neither at the conjunction of filament and buffer layer nor at the interface of electrode and switching layer helps to obtain the self-rectifying resistive memory. Tran et al.[48] proposed a forming-free and self-rectifying unipolar resistive memory based on n+-Si/HfO_x/Ni structure. The memory exhibit excellent self-rectifying behaviour in LRS (>10^3 @ 1 V) and a wide read-out margin for high-density cross-point memory devices. As shown in Figure 3.12a, the typical I-V curve of the DC sweep of n+-Si/HfO_x/Ni shows rectification properties in LRS, which can restrain the crosstalk effect without connecting a diode. And the typical I-V characteristic curves of n+Si/HfO_x/Ni cell in LRS under different working temperatures exhibit excellent self-rectifying characteristics (Rectifying ratio @ 1V is >10^3 at 50 °C) (Figure 3.12b). A model is proposed to clarify the self-rectifying effect in LRS for Ni/HfO_x/n+-Si device. As shown in Figure 3.12c, defects or traps can be introduced into HfO_x/SiO_x dielectrics during the memory programming, whose energy level might align with the mid-gap of the Si substrate. Electrons can be injected from the defect states in HfO_x/SiO_x dielectrics into the Si electrode during reverse bias operation. However, the current would be suppressed due to the Schottky barrier at the n+-Si and HfO_x/SiO_x junction,

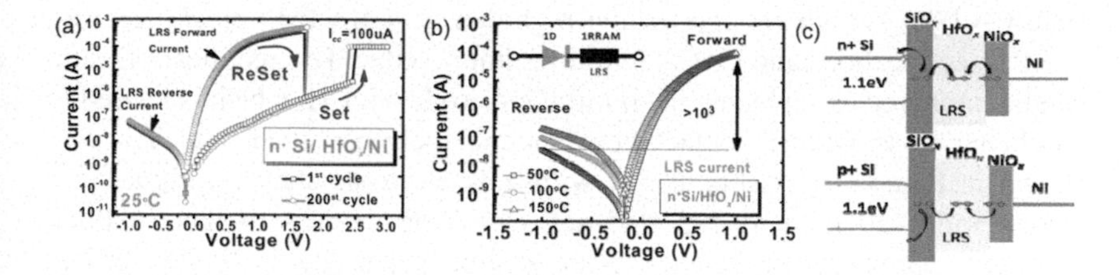

FIGURE 3.12 (a) I-V characteristic curves of n+ Si/HfO_x/Ni cell shows rectification properties in LRS. (b) I-V curves of n+ Si/HfO_x/Ni cell at different working temperature shows excellent self-rectifying behaviour in LRS (>10^3 @ 1 V). (c) The schematic of reverse current transport in n^+-Si/HfO_x/Ni (top) and p^+-Si/HfO_x/Ni (bottom) devices. © [2011] IEEE. Reprinted, with permission, from [X.A. Tran, Self-rectifying and forming-free unipolar HfOx based-high performance RRAM built by fab-avaialbe materials, IEEE Proceedings, and Dec. 1, 2011].

which results in the self-rectifying effect. And such a barrier is not formed in the p^+-Si electrode-based device. But Si-based self-rectifying devices have inherent drawbacks such as high processing temperature, making them incapable of being stacked layer by layer in the BEOL process. Yet inserting a semiconducting buffer layer between the metal electrode and the switching layer can make a stackable self-rectifying device.[49] Lower processing temperature and stackable capability are obtained by introducing the semiconducting material (a-Si). Nevertheless, owing to the high-density defects that exist at the interface between the metal electrode and the oxide layer, the rectifying ratio of the device is decreased to less than 10^2.

Some studies[50,51] report that bilayer structured devices composed of two kinds of CMOS-compatible material show outstanding self-rectifying characteristics. One-layer acts as a forming-free resistive switching (RS) layer and the other layer acts as a rectifier layer with thc help of a highly functional metal. As a well-known high-dielectric material, HfO_2 is attractive in the ReRAM field considering its controllability of resistance values on an atomic scale.[52,53] Ta_2O_5's robust cycling endurance makes it a key role in the ReRAM field.[54] HfO_2 and Ta_2O_5 could have inherently superb uniformity, lacking discrete metallic second phases.[55] However, there has been no study on the self-rectifying properties of these two materials. Yoon et al.[55] proposed a self-rectifying resistance switching memory with the Pt/Ta_2O_5/HfO_2/TiN structure. The HfO_2 layer acts as the resistance switching layer by trapping or de-trapping of electronic carriers, while the Ta_2O_5 layer remains intact during the whole switching cycle and establishes a high Schottky barrier with a high-work-function metal (Pt). As shown in Figure 3.13a, the self-rectifying memory exhibits outstanding characteristics such as being highly uniform and electroforming-free, which is related to the increment of the initial defect (V_o) content in the atomic-layer-deposited HfO_2 layer. In addition, the resistance ratio (~1,000) is high enough for the high-density CBA (Mb block density), but the rectification ratio (10^4) should be improved.[56] The rectification ratio can be improved by replacing the bottom electrode with Ti, which is related to the better quasi-Ohmic contact between the dielectric stack and Ti.[57] As shown in Figure 3.13b, this device with a Ti electrode shows a higher rectifying ratio (10^6). Many reported self-rectifying devices to have a thick rectifier layer, bringing about a poor scaling capability in a 3D vertical structure and large operation voltages, which hinders its application in embedded memory.[55,57–59] Low operation voltage (<3 V) and good compatibility with the CMOS logic devices are achieved by introducing a thin HfO_2 layer (3 nm) between the Pd and a conductive WO_x layer. The high rectifying ratio (>100) caused by Pd/HfO_2 Schottky contact is helpful to suppress the sneaking current (Figure 3.13c). In addition, the device also shows high uniformity and fast operation speed.

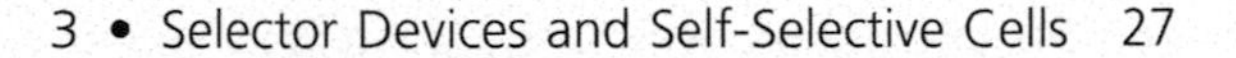

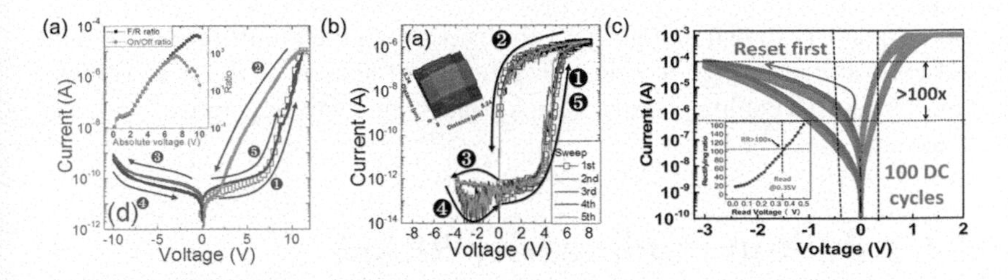

FIGURE 3.13 (a) Resistive switching I-V curves of the device with the Pt/Ta_2O_5/HfO_2/TiN structure. (b) Resistive switching I-V curves of the device with Ti electrode. (c) Typical IV curves of self-rectifying memory with Pd/HfO_2/WO_x/W structure. 3.13(a) © [2014] John Wiley and Sons. Reprinted, with permission, from [Cheol Seong Hwang, Tae Hyung Park, Dae Eun Kwon, et al., Highly Uniform, Electroforming-Free, and Self-Rectifying Resistive Memory in the Pt/Ta_2O_5/HfO_{2-x}/TiN Structure, Advanced Functional Materials, and May 26, 2014]. 3.13(b) © [2015] John Wiley and Sons. Reprinted, with permission, from [Cheol Seong Hwang, Xinglong Shao, Young Jae Kwon, et al., Pt/Ta_2O_5/HfO_2–x/Ti Resistive Switching Memory Competing with Multilevel NAND Flash, Advanced Materials, and May 13, 2015]. 3.13(c) © [2018] IEEE. Reprinted, with permission, from [X.A. Tran, Self-Rectifying and Forming-Free Resistive-Switching Device for Embedded Memory Application, IEEE Electron Device Letters, and May 1, 2018].

The self-rectifying RRAM devices fabricated in 3D vertical structures are critical. Yoon et al.[58] reported a Pt/Ta_2O_5/HfO_2/TiN-based RRAM with a thin switching layer (<20 nm) in a 3D vertical structure (Figure 3.14a). The cross-sectional TEM image showing the overall structure of the deposited HfO_2 layer on the etched BE and SiO_2 layers. The typical self-rectifying

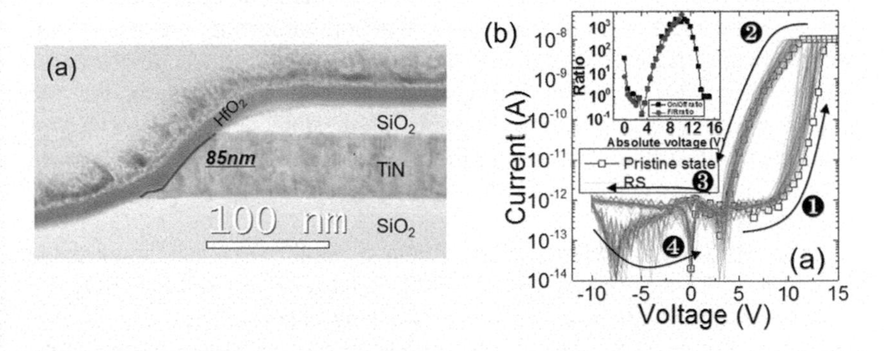

FIGURE 3.14 (a) Cross-sectional TEM image showing the devices fabricated on 3D vertical structure. (b) Resistive switching I-V curves of the vertical Pt/Ta_2O_5/HfO_2/TiN cell.

characteristic is shown in Figure 3.14b. Inset shows the on/off resistance ratio and the forward/reverse rectification ratio as a function of the read voltage. Low current switching (below 10 nA) and high rectifying ratios (10^3) are achieved.

3.2.2 Built-in nonlinearity RRAM

The self-selective cells (SSC) with built-in nonlinearity consist of a selection layer and a memory layer. The selection layer is utilised to offer high non-linearity and the memory layer is used for storage. And the selection layer used in SSC can be divided into two types in light of different electronic transmission mechanisms: threshold type and exponential type.

3.2.2.1 SSC with Threshold Type Selection Layer

Various threshold switching selectors available can be used for constructing SSC, such as insulator-metal transition (IMT) selectors[60,61] and mixed ionic and electron conductor (MIEC)-based selectors.[62] High nonlinearity is offered by the threshold type selection layer in SSC thanks to an abrupt threshold switching I-V characteristic with a hysteresis. The self-selective cells based on insulator-metal transition (IMT) selectors have been investigated. Kim et al.[28] firstly proposed an ultrathin (~10 nm) W/NbO_x/Pt device with both threshold switching (TS) and memory switching (MS) characteristics. Ultrathin Nb_2O_5/NbO_2 stacking with both TS and MS is formed by oxidising NbO_2. A significant reduction of leakage current of the unselected cell ($\pm 1/2V_{read}$ region) emerges in the self-selective cell, compared with RRAM without selector. As shown in Figure 3.15a, excellent characteristics such as high on current (>2 MA/cm^2) and good switching uniformity are also shown in the self-selective cell. In addition, the disturbance of unselect device under read mode (@$\pm 1/2V_{read}$) is negligible (Figure 3.14b). Furthermore, the readout margin was improved by suppressing the LRS current at $1/2V_{read}$ of unselected cell in hybrid memory (Figure 3.14c). However, the IMT-based self-selective cells can be limited due to its high leakage current and low nonlinearity.

The MIEC-based selector embodies high nonlinearity and low-leakage current at off-state compared with the IMT-based selector. And the HfO_2 is a laudable material with salient memory switching characteristics. Luo et al.[63] configured a bilayer SSC device by combining these two promising materials. Outstanding self-selective characteristics are demonstrated within a four-layer VRRAM array, including ultra-low half-select leakage (<0.1 pA), very high nonlinearity (>10^3), low operation current (nA level), self-compliance, high

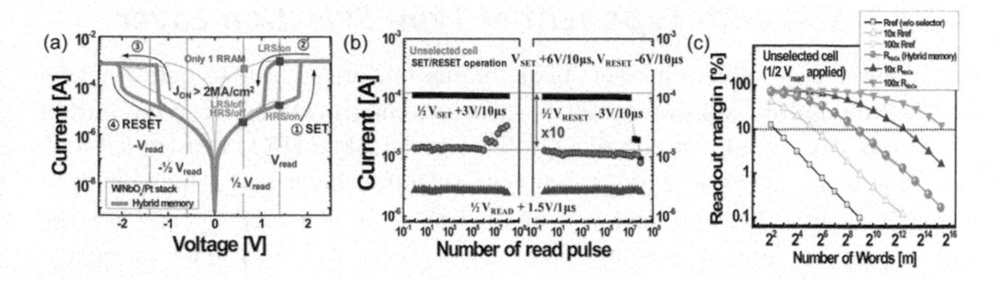

FIGURE 3.15 (a) I-V characteristics of the IMT-based self-selective cell. (b) Disturbance test of unselect device under read mode ($1/2V_{read}$) and set/reset mode ($1/2V_{SET/RESET}$). (c) Calculated readout margin of the IMT based self-selective cell and only 1 RRAM device under the worst case. © [2012] IEEE. Reprinted, with permission, from [Seonghyun Kim, Ultrathin (<10 nm) Nb_2O_5/NbO_2 hybrid memory with both memory and selector characteristics for high density 3D vertically stackable RRAM applications, IEEE Proceedings, and Jun. 1, 2012].

endurance (>10^7), and robust read/write disturbance immunity. The schematic of the four-layer 3D VRRAM array is shown in Figure 3.16a. Typical I-V curve of bilayer SSC shows outstanding characteristics of high selectivity, low operation current, self-compliance, and low leakage current (Figure 3.16b). Inset shows that the programming voltage (V_P) and selective voltage (V_s) are strongly dependent on the thickness of CuGeS. In addition, a sufficient read margin can be maintained in an array of up to 10 Mb in the worst-case condition (Figure 3.16c).

Owning to the movement of ions or local phase change in the selection layer, all the self-selective cells with threshold type selection layer demonstrate a large variation of the turn-on voltage, indicating a limited read voltage window of the 3D VRRAM array.[12,63] Therefore, the uniformity of SSC is critical and needs to be improved.

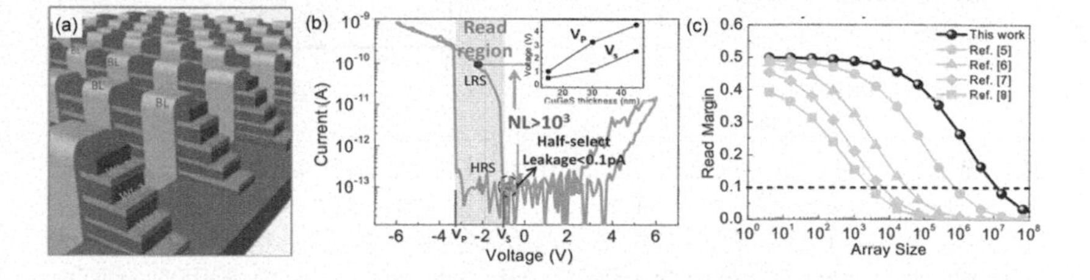

FIGURE 3.16 (a) The schematic of four-layer 3D VRRAM array. (b) Typical I-V curve of bilayer SSC. (c) Read margin analysis in the worst-case condition. © [2015] IEEE. Reprinted, with permission, from [Qing Luo, Demonstration of 3D vertical RRAM with ultra low-leakage, high-selectivity and self-compliance memory cells, IEEE Proceedings, and Dec. 1, 2015].

3.2.2.2 SSC with Exponential Type Selection Layer

Exponential type selection layers boast higher uniformity due to pure electron conduction and no ion movement or structure modification occurring during operation. Chen et al.[51] reported a double-layer stacked HfO_x vertical RRAM, which is made up of an exponential type selection layer and a filament type memory layer. The nonlinear I-V characteristic is achieved by engineering electrode/oxide interface using TiON layer (Figure 3.17a). The fabricated vertical RRAM excels in reset current (<50 μA), switching speed (~50 ns), switching endurance (>10^8 cycles), half-selected read disturbance immunity (>10^9 cycles) and retention (>10^5 s @125 °C). A typical I-V curve is shown

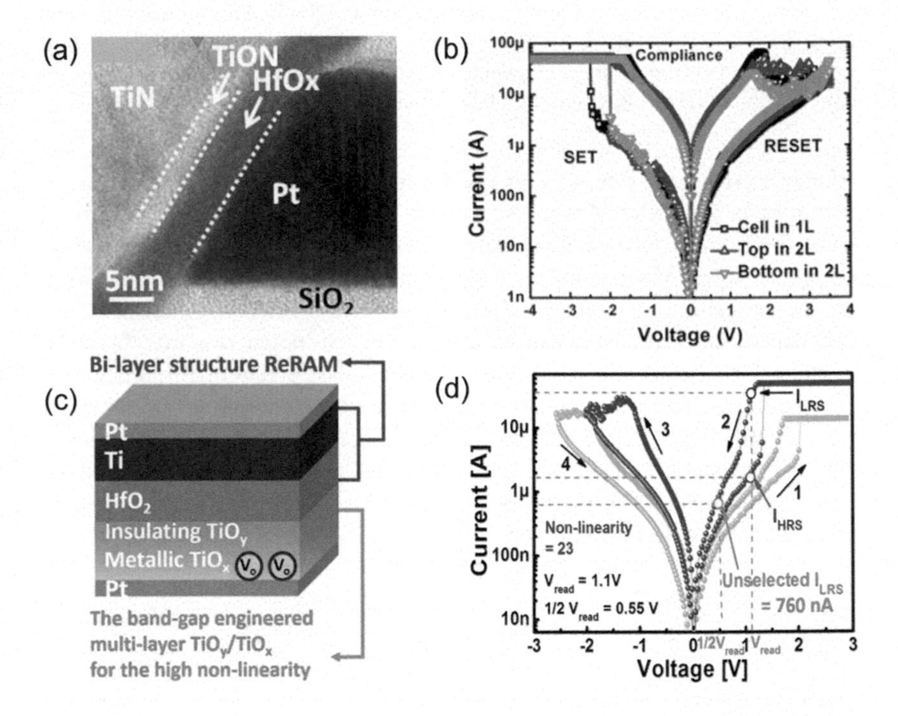

FIGURE 3.17 (a) Cross-sectional TEM image of single-layer sample. (b) Typical DC I-V switching characteristics of bilayer SSC. (c) The schematic of the $Ti/HfO_2/TiO_x/Pt$ device. (d) Typical I-V curves of the $Ti/HfO_2/TiO_x/Pt$ device. 3.17(a, b) © [2012] IEEE. Reprinted, with permission, from [Hong-Yu Chen, HfO_x based vertical resistive random access memory for cost-effective 3D cross-point architecture without cell selector, IEEE Proceedings, and Dec. 1, 2012]. 3.17(c, d) © [2013] IEEE. Reprinted, with permission, from [Sangheon Lee, Selector-less ReRAM with an excellent non-linearity and reliability by the band-gap engineered multi-layer titanium oxide and triangular shaped AC pulse, IEEE Proceedings, and Dec. 1, 2013].

in Figure 3.17b. The formation of an interfacial TiON layer may cause a tunnelling barrier and a large R_{on} for memory. However, the nonlinearity in these devices is low (~5). The nonlinearity can be improved by engineering the band gap of the TiO_y/TiO_x multi-layer.[50] Lee et al.[50] proposed that the nonlinearity and readout margin have been significantly improved by optimising the oxygen profile of the TiO_x layer in the RRAM. The schematic of the $Ti/HfO_2/TiO_x/Pt$ device is shown in Figure 3.17c. The typical I-V curves of the $Ti/HfO_2/TiO_x/Pt$ device reported in this work are vastly nonlinear (~23) (Figure 3.17d). The excellent AC endurance (~10^8 cycles) and switching uniformity also achieved in this device.

However, the SSC with exponential type selection layer mentioned above is also inherently a filament type resistive switching memory, which will cause a large variety of set voltage. Therefore, an SSC with an exponential type selection layer and a non-filament type memory layer is a better choice. Govoreanu et al.[64] put forward an SSC based on $TiN/Al_2O_3/TiO_2/TiN$ structure, where the Al_2O_3 layer plays as an exponential type selection layer and the TiO_2 layer acts as a non-filament type memory layer. The cross-sectional TEM image is shown in Figure 3.18a. The TiO_2 layer is reduced by using a post-TiO_2 deposition anneal at 600 °C, which induces an oxygen vacancy (V_o) profile across the whole film and creates a vacancy modulated conductive oxide (VMCO) active layer. As shown in Figure 3.18b, the VMCO device works so well such as low operation current (1 μA), large on/off window of >10^2, and 100x nonlinearity. The switching mechanism is exhibited in Figure 3.18c, the vacancies are moved back and forth by applying an external bias. The off/initial state resistance originates from the whole tunnelling barrier layer and part of the V_o-depleted VMCO layer, while the on-state resistance is mainly related to the conduction through the tunnelling barrier layer.

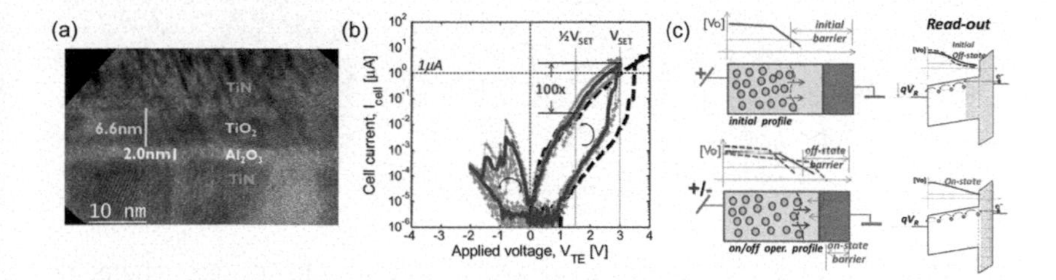

FIGURE 3.18 (a) The cross-sectional TEM image of the VMCO device. (b) Typical DC I-V switching characteristics of the VMCO device. (c) The schematics of the switching model. © [2013] IEEE. Reprinted, with permission, from [B. Govoreanu, Vacancy-modulated conductive oxide resistive RAM (VMCO-RRAM): An area-scalable switching current, self-compliant, highly nonlinear and wide on/off-window resistive switching cell, IEEE Proceedings, and Dec. 1, 2013].

Ma et al.[65] proposed that the non-filament RRAM can be more uniform as its switching is through the uniform defect profile modulation. Govoreanu et al.[66,67] found an a-VMCO nonfilamentary resistive switching memory cell with self-rectifying, self-compliant, forming-free, and analog behaviour. The a-VMCO device is fabricated by using amorphous-Si (a-Si) with larger thickness (8 nm) as a barrier and TiO_2 as a switching layer. The devices are forming-free and initially in the on-state (Figure 3.19a). The VMCO device (Figure 3.19b) has excellent device-to-device uniformity. The barrier resistance modulation helps realise wider on/off window and current reduction while preserving an excellent variability. As shown in Figure 3.19c, the Inner-interface engineered device shows an on/off window above 100 and reset switching currents of down to ~1 μA for 40-nm-size cells, scaling with size, without compromising reliability. However, the nonlinearity and leakage current of the SSC device still need to be improved.

Xu et al.[68] suggested an SSC fabricated on a four-layer vertical structure with a self-alignment technique (Figure 3.20a). The selection layer of this device was TiO_x, which is formed by an oxygen plasma treatment of TiN. The device shows high nonlinearity (>10^3), low leakage current (~0.1 pA), robust endurance, and excellent disturbance immunity. Many reported SSCs[28,50,51,69] are fabricated by directly depositing films into a trench and form a sidewall contact with the stacked word lines (WLs), which may create sneak leakage current path in the selective layer (SL) between adjacent WLs. The inter-layer leakage is successfully eliminated by using a self-alignment technique, extending the scaling limit of VRRAM beyond 5 nm. Typical I-V curves exhibited the hysteresis characteristics (Figure 3.20b). Besides, the nonlinearity

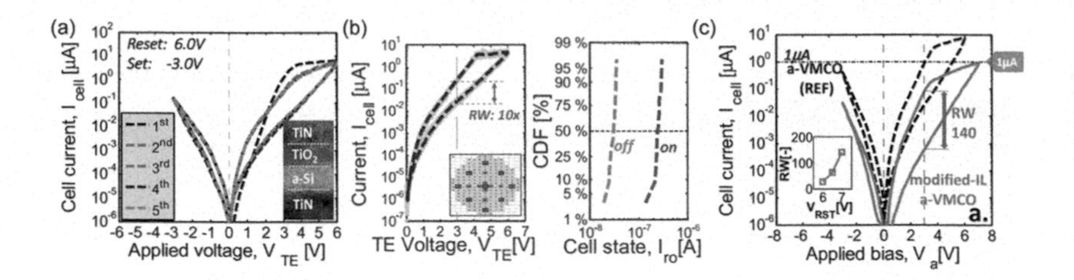

FIGURE 3.19 (a) DC IV sweeps of the a-VMCO device. (b) DC reset characteristics (left) and excellent read-out d2d uniformity (right). (c) DC IV sweeps of the engineered IL a-VMCO device (red) compared with the a-VMCO device (black). 3.19(a)(b) © [2016] IEEE. Reprinted, with permission, from [B. Govoreanu, Advanced a-VMCO resistive switching memory through inner interface engineering with wide (>10^2) on/off window, tunable μA-range switching current and excellent variability, IEEE Proceedings, and Jun. 1, 2016].

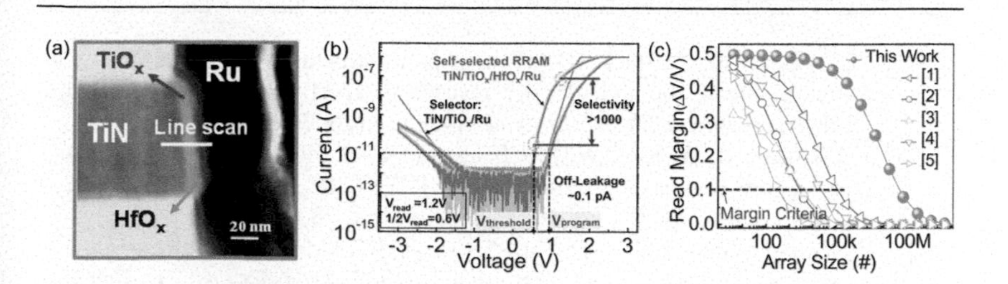

FIGURE 3.20 (a) Magnified TEM image of the TiN/TiO$_x$/HfO TiO$_x$/Ru SSC device. (b) Typical I-V switching characteristics of the TiN/TiO$_x$/HfO$_2$/Ru device. (c) Calculated readout margin upon array size. © [2016] IEEE. Reprinted, with permission, from [Xiaoxin Xu, Fully CMOS compatible 3D vertical RRAM with self-aligned self-selective cell enabling sub 5nm scaling, IEEE Proceedings, and Jun. 1, 2016].

at the low resistance state (LRS) exceeds 10^3, which can get rid of the sneak leakage current in large-scale crossbar arrays. In addition, the calculated readout margin upon array size shows that sufficient read margin can be realised in an array up to 10 Mb (Figure 3.20c). However, there are still many challenges limiting the application of 3D VRRAM even if these SSCs mentioned above perform satisfactorily.

REFERENCES, BIBLIOGRAPHY, OR WORKS CITED

[1] X.P. Wang, et al., "Highly compact 1T-1R architecture (4F2 footprint) involving fully CMOS compatible vertical GAA nano-pillar transistors and oxide-based RRAM cells exhibiting excellent NVM properties and ultra-low power operation." In: *2012 International Electron Devices Meeting* (2012), pp. 20.6.1–20.6.4.

[2] W. Ching-Hua, et al., "Three-dimensional 4F2 ReRAM cell with CMOS logic compatible process." In: *2010 International Electron Devices Meeting* (2010), pp. 29.6.1–29.6.4.

[3] J.H. Oh, et al., "Full integration of highly manufacturable 512Mb PRAM based on 90 nm technology." In: *2006 International Electron Devices Meeting* (2006), pp. 1–4.

[4] Y. Sasago, et al., "Cross-point phase change memory with 4F2 cell size driven by low-contact-resistivity poly-Si diode." In: *2009 Symposium on VLSI Technology* (2009), pp. 24–25.

[5] Y.H. Song, S.Y. Park, J.M. Lee, H.J. Yang, and G.H. Kil, "Bidirectional two-terminal switching device for crossbar array architecture," *IEEE Electron Device Letters*, vol. 32, no. 8, pp. 1023–1025, 2011.

[6] M.J. Lee, et al., "2-stack 1D-1R cross-point structure with oxide diodes as switch elements for high density resistance RAM applications." In: *2007 IEEE International Electron Devices Meeting* (2007), pp. 771–774.
[7] J.-J. Huang, C.-W. Kuo, W.-C. Chang, and T.-H. Hou, "Transition of stable rectification to resistive-switching in Ti/TiO2/Pt oxide diode," *Applied Physics Letters*, vol. 96, no. 26, p. 262901, Jun. 28, 2010.
[8] S. Kim, J. Zhou, and W.D. Lu, "Crossbar RRAM arrays: Selector device requirements during write operation," *IEEE Transactions on Electron Devices*, vol. 61, no. 8, pp. 2820–2826, 2014.
[9] J. Zhou, K. Kim, and W. Lu, "Crossbar RRAM arrays: Selector device requirements during read operation," *IEEE Transactions on Electron Devices*, vol. 61, no. 5, pp. 1369–1376, 2014.
[10] G.W. Burr, et al., "Access devices for 3D crosspoint memory," *Journal of Vacuum Science & Technology B*, vol. 32, no. 4, p. 040802, Jul. 1, 2014/07/01.
[11] Q. Luo, et al., "Fully BEOL compatible TaO_x-based selector with high uniformity and robust performance." In: *2016 IEEE International Electron Devices Meeting (IEDM)* (2016), pp. 11.7.1–11.7.4.
[12] Q. Luo, et al., "Highly uniform and nonlinear selection device based on trapezoidal band structure for high density nano-crossbar memory array," *Nano Research*, vol. 10, no. 10, pp. 3295–3302, Oct. 1, 2017.
[13] J.J. Huang, Y.M. Tseng, C.W. Hsu, and T.H. Hou, "Bipolar nonlinear Ni/TiO_2/Ni selector for 1S1R crossbar array applications," *IEEE Electron Device Letters*, vol. 32, no. 10, pp. 1427–1429, 2011.
[14] W. Lee, et al., "High current density and nonlinearity combination of selection device based on TaO_x/TiO_2/TaO_x structure for one selector–one resistor arrays," *ACS Nano*, vol. 6, no. 9, pp. 8166–8172, Sep. 25, 2012.
[15] E. Cimpoiasu, et al., "Aluminum oxide layers as possible components for layered tunnel barriers," *Journal of Applied Physics*, vol. 96, no. 2, pp. 1088–1093, Jul. 15, 2004.
[16] K.K. Likharev, "Layered tunnel barriers for nonvolatile memory devices," *Applied Physics Letters*, vol. 73, no. 15, pp. 2137–2139, Oct. 12, 1998.
[17] R.J. Malik, T.R. Aucoin, R.L. Ross, K. Board, C.E.C. Wood, and L.F. Eastman, "Planar-doped barriers in GaAs by molecular beam epitaxy," *Electronics Letters*, vol. 16, no. 22, pp. 836–838, 1980.
[18] J.G. Simmons, "Electric tunnel effect between dissimilar electrodes separated by a thin insulating film," *Journal of Applied Physics*, vol. 34, no. 9, pp. 2581–2590, Sep. 1, 1963.
[19] Q. Luo, et al., "Cu BEOL compatible selector with high selectivity (>107), extremely low off-current (~pA) and high endurance (>1010)." In: *2015 IEEE International Electron Devices Meeting (IEDM)*, Washington, DC, USA (2015), pp. 10.4.1–10.4.4.
[20] W. Banerjee, S.H. Kim, S. Lee, S. Lee, D. Lee, and H. Hwang, "Deep insight into steep-slope threshold switching with record selectivity (>4 × 1010) controlled by metal-ion movement through vacancy-induced-percolation path: Quantum-level control of hybrid-filament," *Advanced Functional Materials*, vol. 31, no. 37, p. 2104054, Sep. 1, 2021. 10.1002/adfm.2021 04054

[21] J. Song, J. Woo, A. Prakash, D. Lee, and H. Hwang, "Threshold selector with high selectivity and steep slope for cross-point memory array," *IEEE Electron Device Letters*, vol. 36, no. 7, pp. 681–683, 2015.
[22] B. Grisafe, M. Jerry, J.A. Smith, and S. Datta, "Performance enhancement of Ag/HfO_2 metal ion threshold switch cross-point selectors," *IEEE Electron Device Letters*, vol. 40, no. 10, pp. 1602–1605, 2019.
[23] Q. Hua, et al., "A threshold switching selector based on highly ordered Ag nanodots for X-point memory applications," *Advanced Science*, vol. 6, no. 10, p. 1900024, May 1, 2019. 10.1002/advs.201900024
[24] A. Cavalleri, et al., "Band-selective measurements of electron dynamics in VO_2 using femtosecond near-edge X-ray absorption," *Physical Review Letters*, vol. 95, no. 6, p. 067405, Aug. 5, 2005.
[25] M. Son, et al., "Excellent selector characteristics of nanoscale VO_2 for high-density bipolar ReRAM applications," *IEEE Electron Device Letters*, vol. 32, no. 11, pp. 1579–1581, 2011.
[26] J. Wei, Z. Wang, W. Chen, and D.H. Cobden, "New aspects of the metal–insulator transition in single-domain vanadium dioxide nanobeams," *Nature Nanotechnology*, vol. 4, no. 7, pp. 420–424, Jul. 1, 2009.
[27] J.A.J. Rupp, R. Waser, and D.J. Wouters, "Threshold switching in amorphous Cr-doped vanadium oxide for new crossbar selector." In: *2016 IEEE 8th International Memory Workshop (IMW)* (2016), pp. 1–4.
[28] S. Kim, et al., "Ultrathin (<10 nm) Nb_2O_5/NbO_2 hybrid memory with both memory and selector characteristics for high density 3D vertically stackable RRAM applications." In: *2012 Symposium on VLSI Technology (VLSIT)* (2012), pp. 155–156.
[29] X. Liu, et al., "Diode-less bilayer oxide (WO_x–NbO_x) device for cross-point resistive memory applications," *Nanotechnology*, vol. 22, no. 47, p. 475702, Nov. 4, 2011.
[30] E. Cha, et al., "Nanoscale (~10 nm) 3D vertical ReRAM and NbO_2 threshold selector with TiN electrode." In: *2013 IEEE International Electron Devices Meeting*, Washington, DC, USA (2013), pp. 10.5.1–10.5.4.
[31] J. Park, T. Hadamek, A.B. Posadas, E. Cha, A.A. Demkov, and H. Hwang, "Multi-layered NiO_y/NbO_x/NiO_y fast drift-free threshold switch with high Ion/Ioff ratio for selector application," *Scientific Reports*, vol. 7, no. 1, p. 4068, Jun. 22, 2017.
[32] Q. Luo, et al., "Nb_{1-x} O_2 based universal selector with ultra-high endurance (>1012), high speed (10 ns) and excellent V_{th} stability." In: *2019 Symposium on VLSI Technology*, Kyoto, Japan (2019), pp. T236–T237.
[33] S.R. Ovshinsky, "Reversible electrical switching phenomena in disordered structures," *Physical Review Letters*, vol. 21, no. 20, pp. 1450–1453, Nov. 11, 1968.
[34] H.Y. Cheng, et al., "Si Incorporation into AsSeGe chalcogenides for high thermal stability, high endurance and extremely low V_{th} drift 3D stackable cross-point memory." In: *2020 IEEE Symposium on VLSI Technology*, Honolulu, HI, USA (2020), pp. 1–2.
[35] B. Govoreanu, et al., "Thermally stable integrated Se-based OTS selectors with >20 MA/cm^2 current drive, >3.103 half-bias nonlinearity, tunable threshold voltage and excellent endurance." In: *2017 Symposium on VLSI Technology* (2017), pp. T92–T93.

[36] S. Lee, J. Lee, S. Kim, D. Lee, D. Lee, and H. Hwang, "Mg-Te OTS selector with low I_{off} (<100 pA), fast switching speed (τd = 7 ns), and high thermal stability (400 °C/30 min) for X-point memory applications." In: *2021 Symposium on VLSI Technology*, Kyoto, Japan (2021), pp. 1–2.
[37] S.R. Ovshinsky, "An introduction to ovonic research," *Journal of Non-Crystalline Solids*, vol. 2, pp. 99–106, Jan. 1, 1970.
[38] M. Nardone, V.G. Karpov, D.C.S. Jackson, and I.V. Karpov, "A unified model of nucleation switching," *Applied Physics Letters*, vol. 94, no. 10, p. 103509, Mar. 9, 2009.
[39] N.S. Avasarala, et al., "Half-threshold bias Ioff reduction down to nA range of thermally and electrically stable high-performance integrated OTS selector, obtained by Se enrichment and N-doping of thin GeSe layers." In: *2018 IEEE Symposium on VLSI Technology* (2018), pp. 209–210.
[40] J. Shen, et al., "Elemental electrical switch enabling phase segregation–free operation," *Science*, vol. 374, no. 6573, pp. 1390–1394, Dec. 10, 2021.
[41] J. Yoo, Y. Koo, S.A. Chekol, J. Park, J. Song, and H. Hwang, "Te-based binary OTS selectors with excellent selectivity (>105), endurance (>108) and thermal stability (>450 °C)." In: *2018 IEEE Symposium on VLSI Technology*, Honolulu, HI, USA (2018), pp. 207–208.
[42] D. Garbin, et al., "Composition optimization and device understanding of Si-Ge-As-Te ovonic threshold switch selector with excellent endurance." In: *2019 IEEE International Electron Devices Meeting (IEDM)*, San Francisco, CA, USA (2019), pp. 35.1.1–35.1.4.
[43] K. Yunmo, B. Kyungjoon, and H. Hyunsang, "Te-based amorphous binary OTS device with excellent selector characteristics for x-point memory applications." In: *2016 Symposium on VLSI Technology*, Honolulu, HI, USA (2016), pp. 1–2.
[44] W. Wang, et al., "Enabling universal memory by overcoming the contradictory speed and stability nature of phase-change materials," *Scientific Reports*, vol. 2, no. 1, p. 360, Apr. 11, 2012.
[45] D. Loke, et al., "Breaking the speed limits of phase-change memory," *Science*, vol. 336, no. 6088, pp. 1566–1569, Jun. 22, 2012.
[46] M. Jo, et al., "Novel cross-point resistive switching memory with self-formed schottky barrier." In: *2010 Symposium on VLSI Technology* (2010), pp. 53–54.
[47] E. Linn, R. Rosezin, C. Kügeler, and R. Waser, "Complementary resistive switches for passive nanocrossbar memories," *Nature Materials*, vol. 9, no. 5, pp. 403–406, May 1, 2010.
[48] X.A. Tran, et al., "Self-rectifying and forming-free unipolar HfO_x based-high performance RRAM built by fab-avaialbe materials." In: *2011 International Electron Devices Meeting* (2011), pp. 31.2.1–31.2.4.
[49] H. Lv, Y. Li, Q. Liu, S. Long, L. Li, and M. Liu, "Self-rectifying resistive-switching device with a-Si/WO_3 bilayer," *IEEE Electron Device Letters*, vol. 34, no. 2, pp. 229–231, 2013.
[50] S. Lee, et al., "Selector-less ReRAM with an excellent non-linearity and reliability by the band-gap engineered multi-layer titanium oxide and triangular shaped AC pulse." In: *2013 IEEE International Electron Devices Meeting* (2013), pp. 10.6.1–10.6.4.

[51] H.Y. Chen, S. Yu, B. Gao, P. Huang, J. Kang, and H.S.P. Wong, "HfO_x based vertical resistive random access memory for cost-effective 3D cross-point architecture without cell selector." In: *2012 International Electron Devices Meeting* (2012), pp. 20.7.1–20.7.4.
[52] B. Govoreanu, et al., "10 × 10 nm^2 Hf/HfO_x crossbar resistive RAM with excellent performance, reliability and low-energy operation." In: *2011 International Electron Devices Meeting* (2011), pp. 31.6.1–31.6.4.
[53] P. Gonon, et al., "Resistance switching in HfO_2 metal-insulator-metal devices," *Journal of Applied Physics*, vol. 107, no. 7, p. 074507, Apr. 1, 2010.
[54] M.-J. Lee, et al., "A fast, high-endurance and scalable non-volatile memory device made from asymmetric Ta_2O_{5-x}/TaO_{2-x} bilayer structures," *Nature Materials*, vol. 10, no. 8, pp. 625–630, Aug. 1, 2011.
[55] J.H. Yoon, et al., "Highly uniform, Electroforming-Free, and Self-Rectifying Resistive Memory in the Pt/Ta2O5/HfO2-x/TiN Structure," *Advanced Functional Materials*, vol. 24, no. 32, pp. 5086–5095, Aug. 1, 2014. 10.1002/adfm.201400064
[56] G.H. Kim, et al., "32 × 32 crossbar array resistive memory composed of a stacked Schottky diode and unipolar resistive memory," *Advanced Functional Materials*, vol. 23, no. 11, pp. 1440–1449, Mar. 20, 2013. 10.1002/adfm.201202170
[57] J.H. Yoon, et al., "Pt/Ta2O5/HfO2−x/Ti resistive switching memory competing with multilevel NAND flash," *Advanced Materials*, vol. 27, no. 25, pp. 3811–3816, Jul. 1, 2015. 10.1002/adma.201501167
[58] J.H. Yoon, et al., "Uniform self-rectifying resistive switching behavior via preformed conducting paths in a vertical-type Ta_2O_5/HfO_2–x structure with a sub-μm^2 cell area," *ACS Applied Materials & Interfaces*, vol. 8, no. 28, pp. 18215–18221, Jul. 20, 2016.
[59] K.M. Kim, et al., "Low-power, self-rectifying, and forming-free memristor with an asymmetric programing voltage for a high-density crossbar application," *Nano Letters*, vol. 16, no. 11, pp. 6724–6732, Nov. 9, 2016.
[60] Y. Yang, J. Lee, S. Lee, C.-H. Liu, Z. Zhong, and W. Lu, "Oxide resistive memory with functionalized graphene as built-in selector element," *Advanced Materials*, vol. 26, no. 22, pp. 3693–3699, Jun. 1, 2014. 10.1002/adma.201400270
[61] S.G. Kim, et al., "Improvement of characteristics of NbO_2 selector and full integration of $4F^2$ 2x-nm tech 1S1R ReRAM." In: *2015 IEEE International Electron Devices Meeting (IEDM)*, Washington, DC, USA (2015), pp. 10.3.1–10.3.4.
[62] G.W. Burr, et al., "Large-scale (512 kbit) integration of multilayer-ready access-devices based on mixed-ionic-electronic-conduction (MIEC) at 100% yield." In: *2012 Symposium on VLSI Technology (VLSIT)* (2012), pp. 41–42.
[63] Q. Luo, et al., "Demonstration of 3D vertical RRAM with ultra low-leakage, high-selectivity and self-compliance memory cells." In: *2015 IEEE International Electron Devices Meeting (IEDM)* (2015), pp. 10.2.1–10.2.4.
[64] B. Govoreanu et al., "Vacancy-modulated conductive oxide resistive RAM (VMCO-RRAM): An area-scalable switching current, self-compliant, highly nonlinear and wide on/off-window resistive switching cell." In: *2013 IEEE International Electron Devices Meeting* (2013), pp. 10.2.1–10.2.4.

[65] J. Ma, et al., "Identify the critical regions and switching/failure mechanisms in non-filamentary RRAM (a-VMCO) by RTN and CVS techniques for memory window improvement." In: *2016 IEEE International Electron Devices Meeting (IEDM)* (2016), pp. 21.4.1–21.4.4.
[66] B. Govoreanu, et al., "A-VMCO: A novel forming-free, self-rectifying, analog memory cell with low-current operation, nonfilamentary switching and excellent variability." In: *2015 Symposium on VLSI Technology (VLSI Technology)* (2015), pp. T132–T133.
[67] B. Govoreanu, et al., "Advanced a-VMCO resistive switching memory through inner interface engineering with wide (>102) on/off window, tunable μA-range switching current and excellent variability." In: *2016 IEEE Symposium on VLSI Technology* (2016), pp. 1–2.
[68] X. Xiaoxin, et al., "Fully CMOS compatible 3D vertical RRAM with self-aligned self-selective cell enabling sub-5nm scaling." In: *2016 IEEE Symposium on VLSI Technology* (2016), pp. 1–2.
[69] S.G. Park, et al., "A non-linear ReRAM cell with sub-1μA ultralow operating current for high density vertical resistive memory (VRRAM)." In: *2012 International Electron Devices Meeting* (2012), pp. 20.8.1–20.8.4.

Integration of 3D RRAM

4

Qing Luo

Contents

In order to extend Moore's law, the use of new nonvolatile storage technology becomes inevitable. New nonvolatile memories, including ferroelectric memory, phase change memory, resistive random access memory, and magnetoresistive memory, have received extensive attention. Among them, resistive random access memory (RRAM) is one of the most promising candidates for next-generation nonvolatile memory applications owing to its superior characteristics, including its simple structure, high switching speed, low power consumption, easy 3D stackable integration, and compatibility with standard complementary metal oxide semiconductor (CMOS) process. To achieve large-scale, high-density integration of RRAM, the 3D structure is undoubtedly the ideal choice.

Among all the integrated structures of RRAM, the structure of 1T1R is the most stable, but the 3D integrated process has not found a practical solution. Until 2020, Researchers from CEA-Leti create the first 3D RRAM with a 1T1R structure[1] by using an architecture similar to the nanosheet transistor to achieve a 3D stack of transistors and integrate the RRAM cells at the drain of the transistors, This 3D 1T1R architecture is derived from their GAA CMOS structure and process flow[1]; the main difference is that each horizontal GAA channel features an independent source connected to a bit line (BL) and a drain directly connected to a pillar of RRAM memory cells (Figure 4.1). This includes a horizontal word line (WL) and a vertical bit line/source line (BL/SL). However, this kind of 3D integrated process has high process complexity.

DOI: 10.1201/9781003391586-4

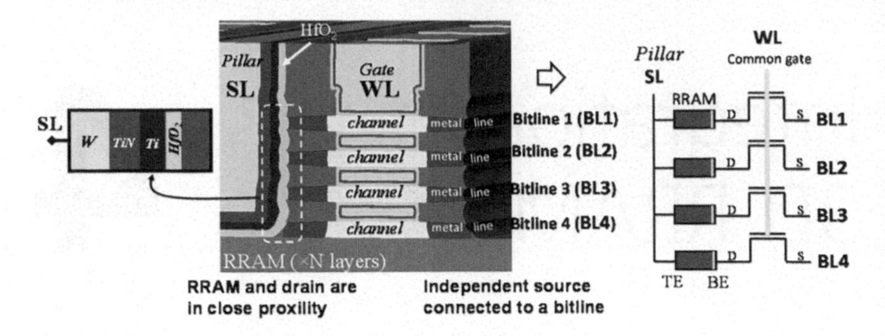

FIGURE 4.1 Cross-sectional schematic of 1T1R O_xRAM memory cell based on GAA nanosheet transistors with an independent bit line (BL) source and a common gate for word line (WL). A 3D pillar is used to connect the source line (SL). © [2020] IEEE. Reprinted, with permission, from [S. Barraud, 3D RRAMs with Gate-All-Around Stacked Nanosheet Transistors for In-Memory-Computing, IEEE Proceedings, and Dec. 12, 2020].

In a crossbar array, the memory cells are located at the cross-point of perpendicularly placed parallel WLs and BLs. In a 2D planer structure, its smallest cell size stands at $4F^2$ (F is the feature size). The cell size per bit will be as small as $4F^2/N$, given an N-layered crossbar array. Compared with the 3D integration of the 1T1R structure, the 3D crossbar array has lower process complexity and higher integration density.

The crossbar array has two forms of 3D integration structure. It can either be stacked layer by layer or vertically placed.[2–5] The former can be easily realised by stacking the planer crossbar array in the vertical direction. The stacked crossbar has several advantages. First, each stack of the crossbar array can be fabricated along with the interconnection process. Second, a separate selector device could be connected in series to the RRAM cell, which makes it possible to optimise the performance of memory and selector individually. Baek et al. of Samsung Company first reported the double-layer crossbar array at the IEDM conference in 2005. They successfully integrated a two-layer test array with a 4 × 5 cross-point matrix. However, this crossbar array did not include a selector device such as a diode or selector. In 2007, Samsung reported a two-layer 8 × 8 3D RRAM with 1D1R (one diode-one resistor) structure at IEDM.[6] P-CuO_x/n-$InZnO_x$ heterojunction thin film was used as an oxide diode which shows high current density (10^4 A/cm^2). And Ti-doped NiO was used for the storage node. The cell size of this work is 0.5 × 0.5 μm. The 3D RRAM array of advanced process nodes needs further exploration. The year 2013 witnessed the emergence of a new cross-point 3D Via RRAM which was successfully demonstrated in a pure 28-nm HKMG CMOS logic process at IEDM. RRAM cell with a cell size of 30 × 30 nm is

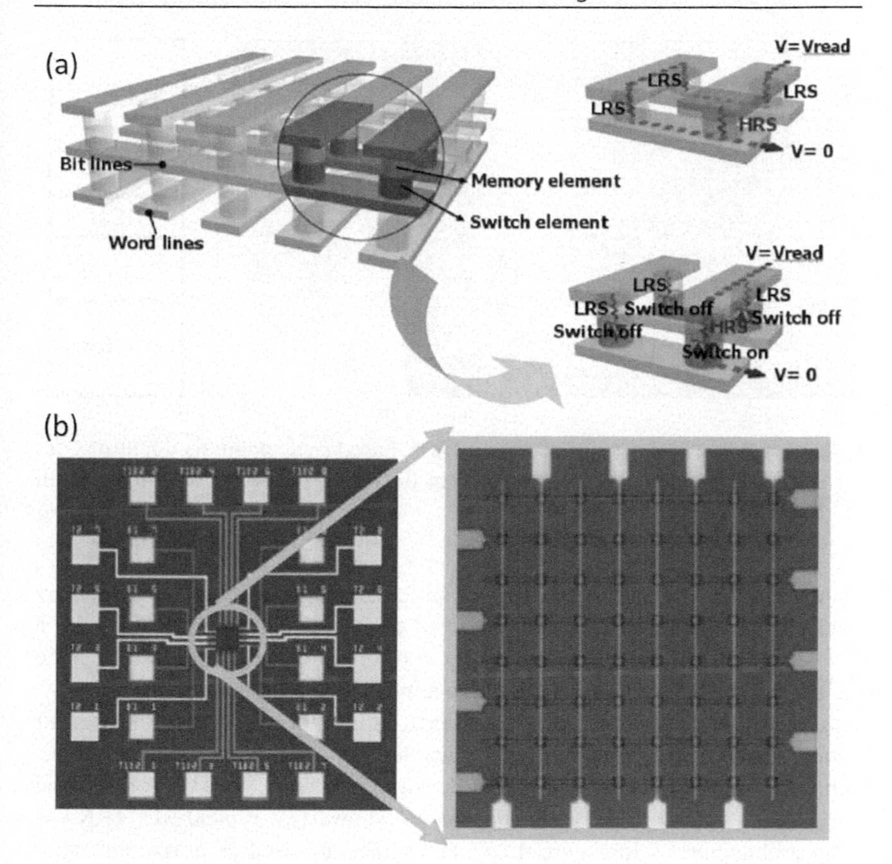

FIGURE 4.2 (a) Generalised cross-point memory structure whose one-bit cell of the array consists of a memory element and a switch element between conductive lines on top (word line) and bottom (bit line). (b) Optical microscopic image including pads for our test array structures and image of the 8 × 8 cell array.[6] © [2007] IEEE. Reprinted, with permission, from [Myoung-Jae Lee, 2-stack 1D-1R Cross-point Structure with Oxide Diodes as Switch Elements for High Density Resistance RAM Applications, IEEE Proceedings, and Dec. 1, 2007].

formed between Cu Via and the landed Cu metal layer in a 28-nm single damascene process, as shown in Figure 4.3. A fully CMOS-compatible TaO_x diode is used to realise the cross-point operation of the 3D crossbar. Sandisk and Toshiba demonstrated a 32-Gb ReRAM[7] prototype in ISSCC at a 24-nm node, representing a major increase in density compared with prior ReRAM developments.

Compared with other stacked structures, the 3D VRRAM has several advantages. Firstly, in sidewall devices, at least one active device dimension

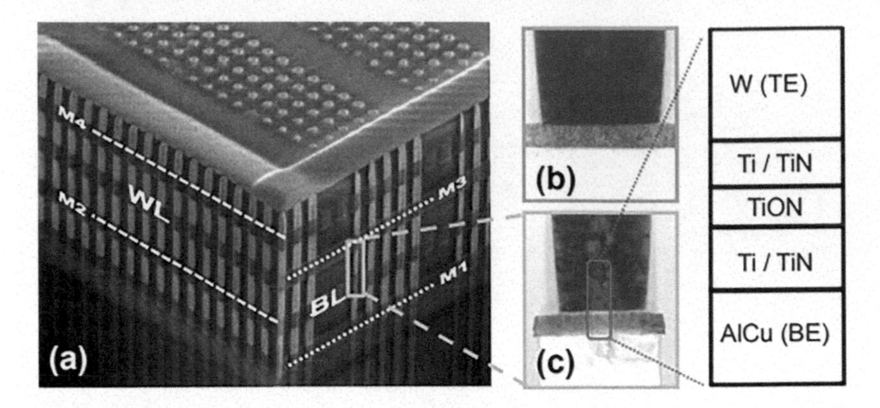

FIGURE 4.3 (a) SEM picture of 28 nm TaON-based cross-point 3D Via RRAM. (b) TEM picture of a standard Cu Via. (c) TEM picture of Via RRAM cell with 30 × 30 nm cell size.[7]

is not critically dependent on lithography. The top electrode is defined by lithography, and the bottom electrode is determined by the thickness of a deposited film, which can be precisely controlled to the atomic level. Since the deposition thickness can be a much smaller size than defined lithography, device-to-device variation can be improved. Furthermore, the number of critical lithography steps will not increase as the stacking layers rise,[5] implying much lower cost and higher density. Yoon et al.[8] firstly proposed the 3D VRRAM structure in 2009. Figure 4.4 shows two typical 3D VRRAM array architectures.[9] In Figure 4.2a, metal lines are used as horizontal word lines, while in Figure 4.2b metal planes are used as horizontal word lines. And both of them use vertical pillars as BLs. In 2011, Baek et al.[10] demonstrated the metal line-based 3D VRRAM. Figure 4.5 shows the process flow of this structure: (a) Multi-layered silicon oxide and silicon nitride are deposited; (b, c) deep hole etching and a switching layer/vertical metal electrode deposition; (d) after the stacked layers are etched, the silicon nitride layers then are removed by using wet-etching; (e) the remaining space is filled with metal to form a horizontal line; (f) the new deposited metal is patterned to form the horizontal line electrode. The metal line-based 3D VRRAM is twice as dense as the metal plane-based 3D VRRAM because both the left and right sides of the metal pillar can form an RRAM cell. However, it is more vulnerable to serious IR drop and RC delay effects.

In 2012, Chen et al.[11] demonstrated a metal plane-based 3D VRRAM. In this structure, only the first pattern (forming a deep hole) is a critical lithographic step. So in this way, a cost-effective 3D RRAM structure was proposed. Due to the etching process in their university lab, the trench is not perfectly

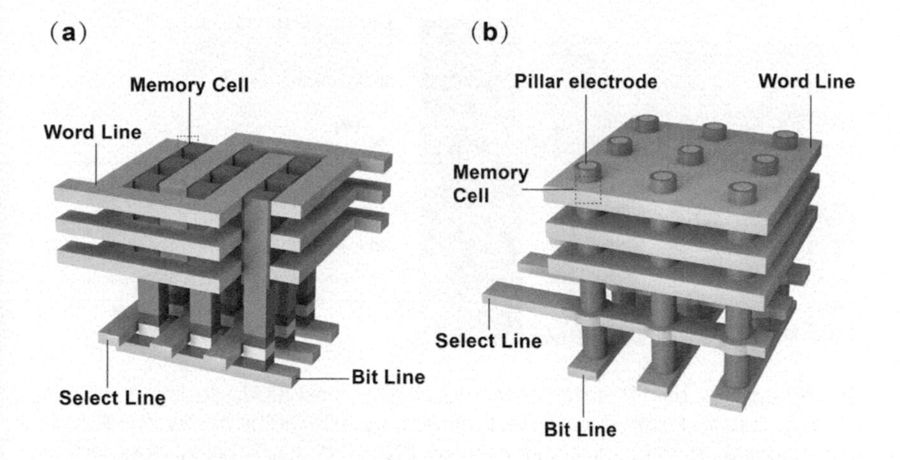

FIGURE 4.4 Two typical 3D VRRAM array architectures. (a) The memory cell sits between the horizontal word line and the vertical pillar. (b) The memory cell sits between the plane electrode and the vertical pillar. © [2011] IEEE. Reprinted, with permission, from [I.G. Baek, Realization of vertical resistive memory (VRRAM) using cost effective 3D process, IEEE Proceedings, and Dec. 1, 2011].

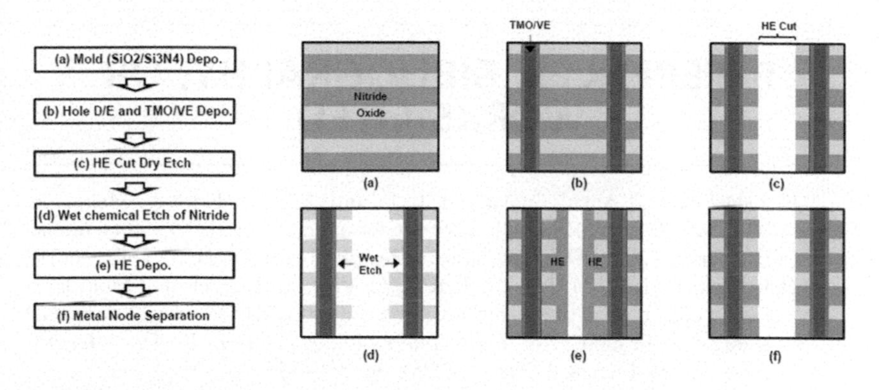

FIGURE 4.5 Key process flow of the metal line-based 3D VRRAM. © [2013] IEEE. Reprinted, with permission, from [Yexin Deng, Design and optimization methodology for 3D RRAM arrays, IEEE Proceedings, and Dec. 1, 2013].

vertical. But it is not a fundamental issue of the device structure. Yue Bai et al.[12] have proposed a much better vertical structure. Several 3D VRRAMs have also been proposed with different horizontal plane electrodes such as W,[13] Ti,[14,15] and TiN.[16–18] Luo et al. invented the four-layer 3D VRRAM in 2015.[19] And in 2017, they proposed an eight-layer 3D VRRAM, as shown in Figure 4.6.[20] The cost of critical masks for metal lines and via contacts that made a stacked 3D

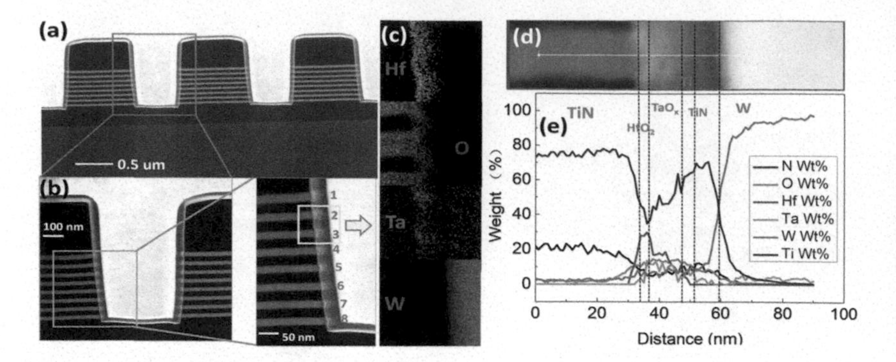

FIGURE 4.6 (a, b) TEM image of the novel 3D vertical RRAM structure. 500-nm hole structure and eight layers of vertical memory cells can be observed clearly. (c) EELS mapping. (d, e) EDX line scan over the TiN/HfO_2/TaO_x/Ti/TiN/W cross-section.

RRAM become unaffordable when the number of stacked layers exceeds eight.[10] Thus the eight-layer work can experimentally prove the advantages of 3D VRRAM, compared with stacked 3D RRAM.

REFERENCES, BIBLIOGRAPHY, OR WORKS CITED

[1] T. Dubreuil, P. Amari, S. Barraud, J. Lacord, E. Esmanhotto, V. Meli, et al. "A novel 3D 1T1R RRAM architecture for memory-centric hyperdimensional computing." *2022 IEEE International Memory Workshop (IMW)*. IEEE (2022).

[2] I. Baek, D. Kim, M. Lee, H.-J. Kim, E. Yim, M. Lee, et al., "Multi-layer cross-point binary oxide resistive memory (OxRRAM) for post-NAND storage application." In: *IEEE International Electron Devices Meeting*, 2005. IEDM Technical Digest (2005), pp. 750–753.

[3] M.-J. Lee, Y. Park, B.-S. Kang, S.-E. Ahn, C. Lee, K. Kim, et al., "2-stack 1D-1R cross-point structure with oxide diodes as switch elements for high density resistance RAM applications." In: *2007 IEEE International Electron Devices Meeting* (2007), pp. 771–774.

[4] I. Baek, C. Park, H. Ju, D. Seong, H. Ahn, J. Kim, et al., "Realization of vertical resistive memory (VRRAM) using cost-effective 3D process." In: *2011 International Electron Devices Meeting* (2011), pp. 31.8.1–31.8.4.

[5] S.-G. Park, M.K. Yang, H. Ju, D.-J. Seong, J.M. Lee, E. Kim, et al., "A non-linear ReRAM cell with sub-1μA ultralow operating current for high density vertical resistive memory (VRRAM)." In: *2012 International Electron Devices Meeting* (2012), pp. 20.8.1–20.8.4.

[6] M.-J. Lee, Y. Park, B.-S. Kang, S.-E. Ahn, C. Lee, K. Kim, et al., "2-stack 1D-1R cross-point structure with oxide diodes as switch elements for high density resistance RAM applications." *2007 IEEE International Electron Devices Meeting*. IEEE (2007).
[7] T.-y. Liu, T. Yan, R. Scheuerlein, Y. Chen, J. Lee, G. Balakrishnan, et al., "A 130.7-mm^2 2-Layer 32-Gb ReRAM memory device in 24-nm technology," *IEEE Journal of Solid-State Circuits*, vol. 49.1, pp. 140–153, 2013.
[8] H.S. Yoon, I.-G. Baek, J. Zhao, H. Sim, M.Y. Park, H. Lee, et al., "Vertical cross-point resistance change memory for ultra-high density non-volatile memory applications." In: *2009 Symposium on VLSI Technology* (2009), pp. 26–27.
[9] Y. Deng, H.-Y. Chen, B. Gao, S. Yu, S.-C. Wu, L. Zhao, et al., "Design and optimization methodology for 3D RRAM arrays." In: *2013 IEEE International Electron Devices Meeting* (2013), pp. 25.7.1–25.7.4.
[10] I. Baek, C. Park, H. Ju, D. Seong, H. Ahn, J. Kim, et al., "Realization of vertical resistive memory (VRRAM) using cost effective 3D process." In: *2011 International Electron Devices Meeting* (2011), pp. 31.8.1–31.8.4.
[11] H.-Y. Chen, S. Yu, B. Gao, P. Huang, J. Kang, and H.-S.P. Wong, "HfOx based vertical resistive random access memory for cost-effective 3D cross-point architecture without cell selector." In: *2012 International Electron Devices Meeting* (2012), pp. 20.7.1–20.7.4.
[12] Y. Bai, H. Wu, R. Wu, Y. Zhang, N. Deng, Z. Yu, et al., "Study of multi-level characteristics for 3D vertical resistive switching memory," *Scientific Reports*, vol. 4, p. 5780, 2014.
[13] S. Gaba, P. Sheridan, C. Du, and W. Lu, "3-D vertical dual-layer oxide memristive devices," *IEEE Transactions on Electron Devices*, vol. 61, pp. 2581–2583, 2014.
[14] C.-W. Hsu, C.-C. Wan, I.-T. Wang, M.-C. Chen, C.-L. Lo, Y.-J. Lee, et al., "3D vertical TaO x/TiO_2 RRAM with over 10^3 self-rectifying ratio and sub-μA operating current." In: *2013 IEEE International Electron Devices Meeting* (2013), pp. 10.4.1–10.4.4.
[15] I.-T. Wang, Y.-C. Lin, Y.-F. Wang, C.-W. Hsu, and T.-H. Hou, "3D synaptic architecture with ultralow sub-10 fJ energy per spike for neuromorphic computation." In: *2014 IEEE International Electron Devices Meeting* (2014), pp. 28.5.1–28.5.4.
[16] H. Li, K.-S. Li, C.-H. Lin, J.-L. Hsu, W.-C. Chiu, M.-C. Chen, et al., "Four-layer 3D vertical RRAM integrated with FinFET as a versatile computing unit for brain-inspired cognitive information processing." In: *2016 IEEE Symposium on VLSI Technology* (2016), pp. 1–2.
[17] H. Li, T.F. Wu, A. Rahimi, K.-S. Li, M. Rusch, C.-H. Lin, et al., "Hyperdimensional computing with 3D VRRAM in-memory kernels: Device-architecture co-design for energy-efficient, error-resilient language recognition." In: *2016 IEEE International Electron Devices Meeting (IEDM)* (2016), pp. 16.1.1–16.1.4.
[18] E. Cha, J. Woo, D. Lee, S. Lee, J. Song, Y. Koo, et al., "Nanoscale (~10 nm) 3D vertical ReRAM and NbO_2 threshold selector with TiN electrode." In: *2013 IEEE International Electron Devices Meeting* (2013), pp. 10.5.1–10.5.4.

[19] Q. Luo, X. Xu, H. Liu, H. Lv, T. Gong, S. Long, et al., "Demonstration of 3D vertical RRAM with ultra low-leakage, high-selectivity and self-compliance memory cells." In: *2015 IEEE International Electron Devices Meeting (IEDM)* (2015), pp. 10.2.1–10.2.4.

[20] Q. Luo, X. Xu, T. Gong, H. Lv, D. Dong, H. Ma, et al., "8-layers 3D vertical RRAM with excellent scalability towards storage class memory applications." In: *2017 IEEE International Electron Devices Meeting (IEDM)* (2017), pp. 2.7.1–2.7.4.

Reliability Issues of the 3D Vertical RRAM

Tiancheng Gong and Dengyun Lei

Contents

5.1 RTN-BASED DEFECT TRACKING TECHNIQUE

In this 3D VRRAM, we exploit the RTN-Based Defect Tracking Technique to characterise the profiles of defects. Figure 5.1a shows the time constants of the RTN signal. Note that different time constant variation rates represent different cases (Figure 5.1b). Positive (Figure 5.1c) and negative (Figure 5.1d) time constant variation rates stand for the respective defect reactions with TE and BE. The locations and energy levels of defects can then be calculated by means of eqs. 1 & 2. It should be mentioned that the double-layer structure of non-filament switching devices should be considered when

DOI: 10.1201/9781003391586-5

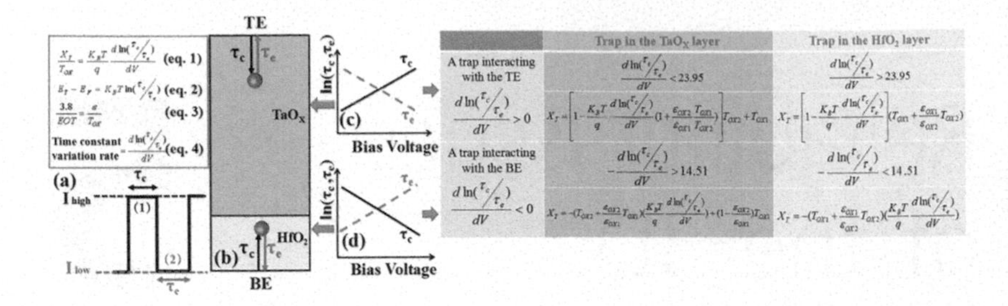

FIGURE 5.1 (a) Definition of the physical parameters. τ_e (emission time) and τ_c (capture time). (b–d) Defect's location and energy level (XT, ET) can be extracted from the capture and emission time constant dependence on the bias VTE using eqs. 1 & 2: (c) Positive or (d) negative time constant variation rate represents that the defect reacts with TE or BE, respectively. Note that the TO_X of each layer should be replaced by Equivalent Oxide Thickness (EOT) using eq. 3. © [2018] IEEE. Reprinted, with permission, from [Tiancheng Gong, Switching and Failure Mechanism of Self-Selective Cell in 3D VRRAM by RTN-Based Defect Tracking Technique, IEEE Proceedings, and May 1, 2018].

developing the defect probing model based on RTN, which is similar to those in MOSFETs with a high-k dielectric layer and a SiO_2 interfacial layer.[1–5] TOX of each layer should be replaced by Equivalent Oxide Thickness (EOT) using eq. 3. Table 5.1 shows all four cases and the corresponding equation to characterise the physical depth of traps. RTN-based defect tracking technique is carried out as follows: The non-filament switching device is switched alternatively between HRS and LRS for several cycles by DC sweeps. After each switch operation, either set or reset, an RTN measurement follows. RTN is measured under stepping TE bias from 2.5 to 2.9 V with $V_{step} = 0.1$ V and $T_{step} = 1$ s. After getting the mean time constants (τ_e and τ_c) of each VTE, we need to determine whether this trap interacts with the TE or BE first. This information can be gained from the polarity of the time constant variation rate.[6] If the polarity is negative, the defect interacts with the BE, and vice versa. The next step is to determine which layer the defect locates in as this TaO_X/HfO_2 device has two layers. The value of the time constant variation rate determines whether a trap is located in the TaO_X layer or the HfO_2 layer. It should be noted that we only consider the time constants of two-level RTN which is caused by one trap, instead of multi-level RTN which is caused by many traps.[7] Figure 5.2 shows one type of testing result of the time constant variation rate being –9.24. The result means this trap locates in the HfO_2 layer and interacts with BE (Figure 5.2b).

TABLE 5.1 Equations characterising the physical depth of traps, ($\varepsilon_{OX1}/\varepsilon_{OX2}$ is the dielectric constant of HfO_2/TaO_X)

	TRAP IN THE TAO_X LAYER	*TRAP IN THE HFO_2 LAYER*
A trap interacting with the TE	$\frac{d\ln(\tau_c/\tau_e)}{dV} < 23.95$	$\frac{d\ln(\tau_c/\tau_e)}{dV} > 23.95$
$\frac{d\ln(\tau_c/\tau_e)}{dV} > 0$	$X_T = \left[1 - \frac{K_BT}{q}\frac{d\ln(\tau_c/\tau_e)}{dV}\left(1 + \frac{\varepsilon_{OX2}}{\varepsilon_{OX1}}\frac{T_{OX1}}{T_{OX2}}\right)\right]T_{OX2} + T_{OX1}$	$X_T = \left[1 - \frac{K_BT}{q}\frac{d\ln(\tau_c/\tau_e)}{dV}\right]\left(T_{OX1} + \frac{\varepsilon_{OX1}}{\varepsilon_{OX2}}T_{OX2}\right)$
A trap interacting with the BE	$-\frac{d\ln(\tau_c/\tau_e)}{dV} > 14.51$	$-\frac{d\ln(\tau_c/\tau_e)}{dV} < 14.51$
$\frac{d\ln(\tau_c/\tau_e)}{dV} < 0$	$X_T = -\left(T_{OX2} + \frac{\varepsilon_{OX2}}{\varepsilon_{OX1}}T_{OX1}\right)\left(\frac{K_BT}{q}\frac{d\ln(\tau_c/\tau_e)}{dV}\right) + \left(1 - \frac{\varepsilon_{OX2}}{\varepsilon_{OX1}}\right)T_{OX1}$	$X_T = -\left(T_{OX1} + \frac{\varepsilon_{OX1}}{\varepsilon_{OX2}}T_{OX2}\right)\left(\frac{K_BT}{q}\frac{d\ln(\tau_c/\tau_e)}{dV}\right)$

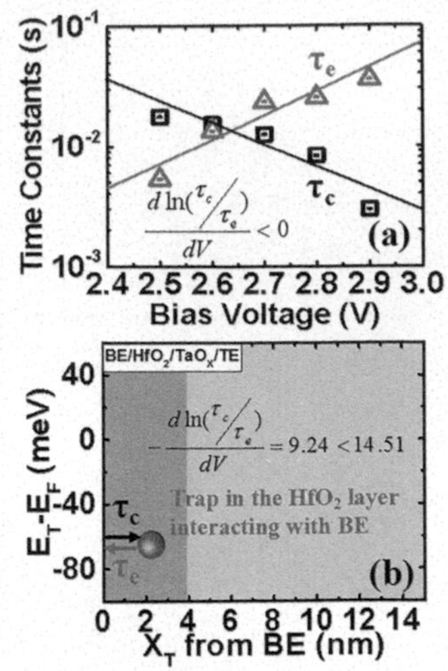

FIGURE 5.2 (a) τ_e and τ_c are dependent on the bias voltage. (b) X_T and E_T of this trap are extracted. Negative time constant variation rate represents that this trap interacts with BE and the value determines that it locates in the HfO_2 layer. © [2018] IEEE. Reprinted, with permission, from [Tiancheng Gong, Switching and Failure Mechanism of Self-Selective Cell in 3D VRRAM by RTN-Based Defect Tracking Technique, IEEE Proceedings, and May 1, 2018].

5.2 SWITCHING MECHANISM OF SELF-SELECTIVE CELL IN 3D VRRAM

Based on the technique mentioned above, defects are detected in both HRS and LRS during normal DC switching cycles, as shown in Figure 5.3a, b. The defect-less region in TaO_X layer in HRS is clearly observed which corresponds to the resistive switching.[8–10] The results show that there are fewer defects in the HfO_2 layer, although many defects in the HfO_2/TaO_X interface. Figure 5.4 is the XPS data of bilayer films in our device, which clearly shows non-stoichiometric TaO_X with metal Ta (leading to the generation of oxygen vacancies) and relatively stoichiometric HfO_2, respectively. This is confirmed by the defect profile in Figure 5.3. The switching mechanism of 3D VRRAM can be understood by the modulation of vacancies under an electric field. ALD-deposited HfO_2 acts as a barrier layer while sputtering-deposited TaO_X acts as a switching layer. Figure 5.8c, d shows the schematics of the switching model: a defect (Vo) profile is induced throughout the whole cell volume of the TaO_X layer. The

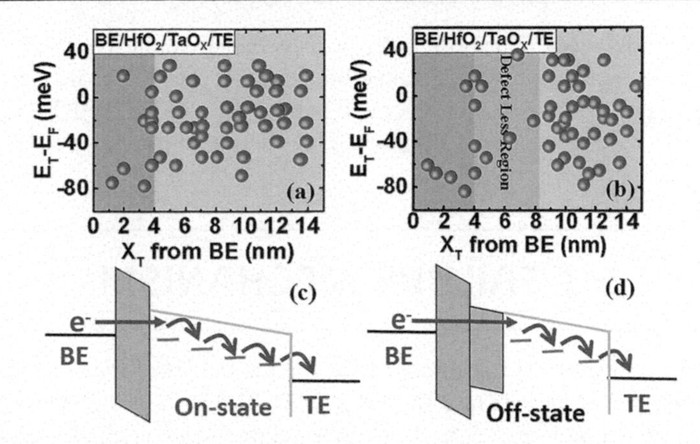

FIGURE 5.3 Defects are detected at (a) LRS and (b) HRS during normal DC switching cycles using the technique in Figures 5.5 and 5.6. Note that in both LRS and HRS, there are fewer defects in the HfO_2 layer, although many defects in the HfO_2/TaO_X interface. There is a defect-less region in the TaO_X layer when the device is switched to LRS. (c, d) Schematic of the switching model: The vacancies are moved back and forth under positive and negative bias. © [2018] IEEE. Reprinted, with permission, from [Tiancheng Gong, Switching and Failure Mechanism of Self-Selective Cell in 3D VRRAM by RTN-Based Defect Tracking Technique, IEEE Proceedings, and May 1, 2018].

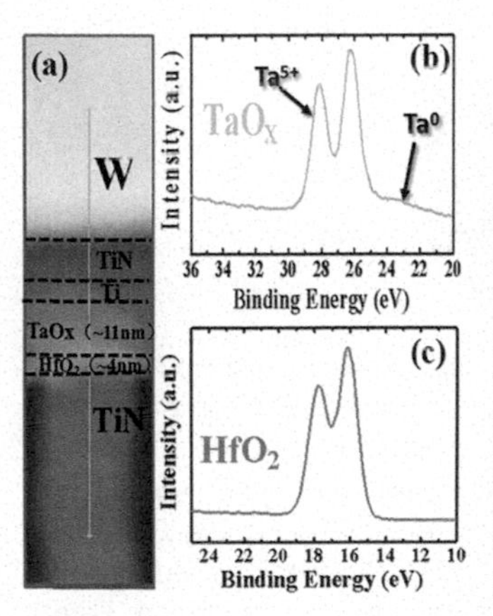

FIGURE 5.4 (a) TEM of this structure XPS spectra of (b) non-stoichiometric TaO_X film indicates that Ta metal exists in the film, which means oxygen vacancies are generated in this layer and (c) relatively stoichiometric HfO_2. © [2018] IEEE. Reprinted, with permission, from [Tiancheng Gong, Switching and Failure Mechanism of Self-Selective Cell in 3D VRRAM by RTN-Based Defect Tracking Technique, IEEE Proceedings, and May 1, 2018].

vacancies move back and forth under positive/negative electric bias, which leads to the adjustment of the tunnelling barrier. The off-state barrier is given by the tunnelling barrier layer and part of the Vo-depleted TaO_X layer and the on-state resistance is dominated by the tunnelling barrier.[11–15]

5.3 FAILURE MECHANISM

Reliability is one of the essential criteria for the universal memory application. In the DC endurance test, the HfO_2/TaO_X device is switched for 1,000 cycles, as shown in Figure 5.5a. It is observed that R_{HRS} gradually decreases while R_{LRS} keeps constant. RTN measurement is carried out in both LRS (*A*, *B*) and HRS (*A*′, *B*′). The defect profiles of B′ is shown in Figure 5.5b, which shows that similar to previous results, the defect-less region is much wider at HRS than at LRS. During cycling, when the R_{HRS} declines, the width of the defect-less region decreases correspondingly (Figure 5.5c). In the retention test, R_{LRS} gradually increases while R_{HRS} remains unchanged when baking at 125 °C for 10,000s, as shown in Figure 5.6a. This can be explained by the thickening of the defect-less region in contrast with the endurance test (Figure 5.6b, c). In conclusion, by monitoring the defect-less

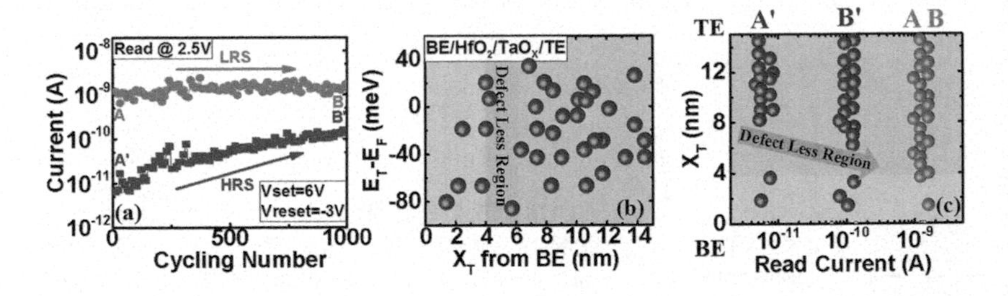

FIGURE 5.5 (a) DC endurance of 1,000 cycles. R_{LRS} keeps constant as cycling number increases while R_{HRS} decreases. (b) Defect profiles of *B*′. Compared with defect profiles of *A*′ (Figure 5.8b), defect-less region becomes narrower. (c) Defect profiles of *A*, *A*′, *B*, and *B*′. © [2018] IEEE. Reprinted, with permission, from [Tiancheng Gong, Switching and Failure Mechanism of Self-Selective Cell in 3D VRRAM by RTN-Based Defect Tracking Technique, IEEE Proceedings, and May 1, 2018]. © [2018] IEEE. Reprinted, with permission, from [Tiancheng Gong, Switching and Failure Mechanism of Self-Selective Cell in 3D VRRAM by RTN-Based Defect Tracking Technique, IEEE Proceedings, and May 1, 2018].

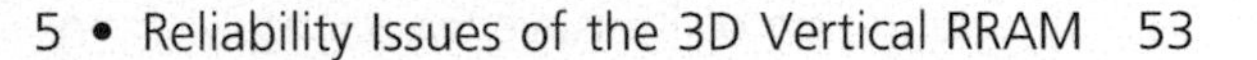

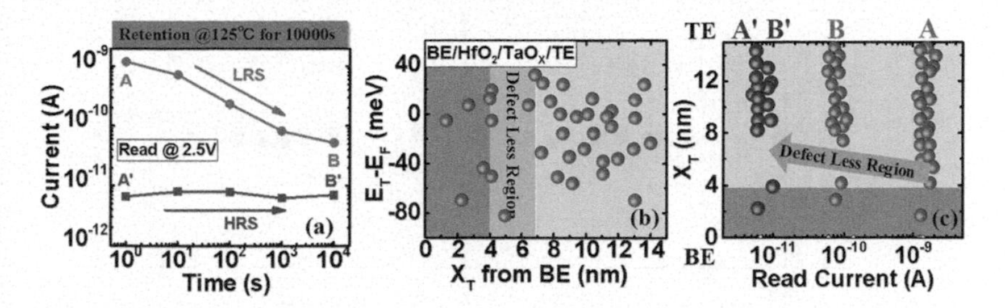

FIGURE 5.6 (a) Retention test for 10^4 s baking at 125 °C. Note that when baking, R_{HRS} stays the same while R_{LRS} becomes larger. (b) Defect profiles of *B*. (c) Defect profiles of *A*, *A'*, *B*, and *B'*. © [2018] IEEE. Reprinted, with permission, from [Tiancheng Gong, Switching and Failure Mechanism of Self-Selective Cell in 3D VRRAM by RTN-Based Defect Tracking Technique, IEEE Proceedings, and May 1, 2018].

region in different cycling numbers and retention times, the main reason for the on/off ratio degradation was revealed.

5.4 MOISTURE STRESS

Moisture is a common type of environmental stress.[16–18] Under 85°C/85% relative humidity (RH) conditions, the HfO_2/TaO_X device is stressed shown in Figure 5.7a–d. It is shown the μ_RHRS declines by 0.36 decades after 61.67 hours of stress, while μ_RLRS shows only a fluctuation within ±0.034 decades referring to the fresh value. As the difference between μ_RHRS and μ_RLRS, μ_Window manifests a 0.39-decade decrease compared to a fresh state, which might cause failure in RRAM memory circuits. Moreover, the degradation hardly recovers under 105°C 24 hours of baking, which makes the reliability concern more severe in humid environments. Figure 5.8 shows the I-V characteristics of VRRAM under 85°C/85%. The stress increases the reset current at the negative bias region which increases the sneak path current.[19–23] Based on Poole-Frenkel's emission, the resistance is a positive correlation to the tunnel barrier. The moisture invasion HfO_2/TaO_x interface along the sidewall induces defects in defect-less region and reduces the tunnel barrier of the interface.[24,25] Tunnel barrier drop causes the degradation of RHRS.

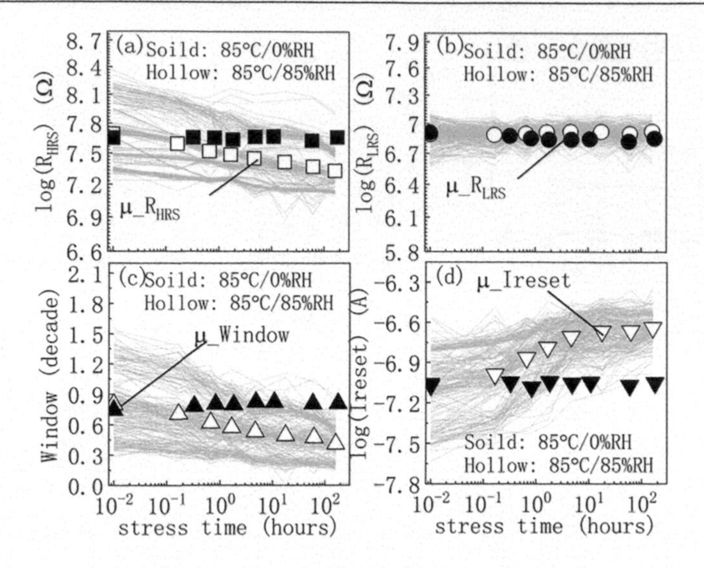

FIGURE 5.7 Aging kinetics of (a) R_{HRS}, (b) R_{LRS}, (c) Window, and (d) I_{reset} on multiple VRRAM devices under 85°C/85% RH moisture stress. © [2020] IEEE. Reprinted, with permission, from [Rui Gao, Effect of Moisture Stress on the Resistance of HfO_2/TaO_x-Based 8-Layer 3D Vertical Resistive Random Access Memory, IEEE Electron Device Letters, and Jan. 1, 2020].

5.5 PROCESS DEVIATION

In VRRAM, HfO_2/TaO_x bilayer is deposited on the sidewall. Figure 5.9 shows the resistance statistics of RLRS and RHRS across eight layers. Both RLRS and RHRS follow a log-normal distribution. μ_RLRS and σ_RLRS are layer-independent while μ_RHRS and σ_RHRS decline from the top layer to the bottom layer.[26] A successive length increase is observed from the high-resolution transmission electron microscope (HRTEM) image, as shown in Figure 5.10. Under the same SET/RESET voltage, the top layer devices suffer from a higher electric field compared to the lower layers, resulting in a fiercer sweep of traps and a wider defect-less region after reset operation, eventually giving rise to a wider barrier and higher resistance at HRS.[26]

In VRRAM, HfO_2/TaO_x bilayer is deposited on the sidewall. Figure 5.10 shows the resistance statistics of RLRS and RHRS across eight layers. Both RLRS and RHRS follow a log-normal distribution. μ_RLRS and σ_RLRS are

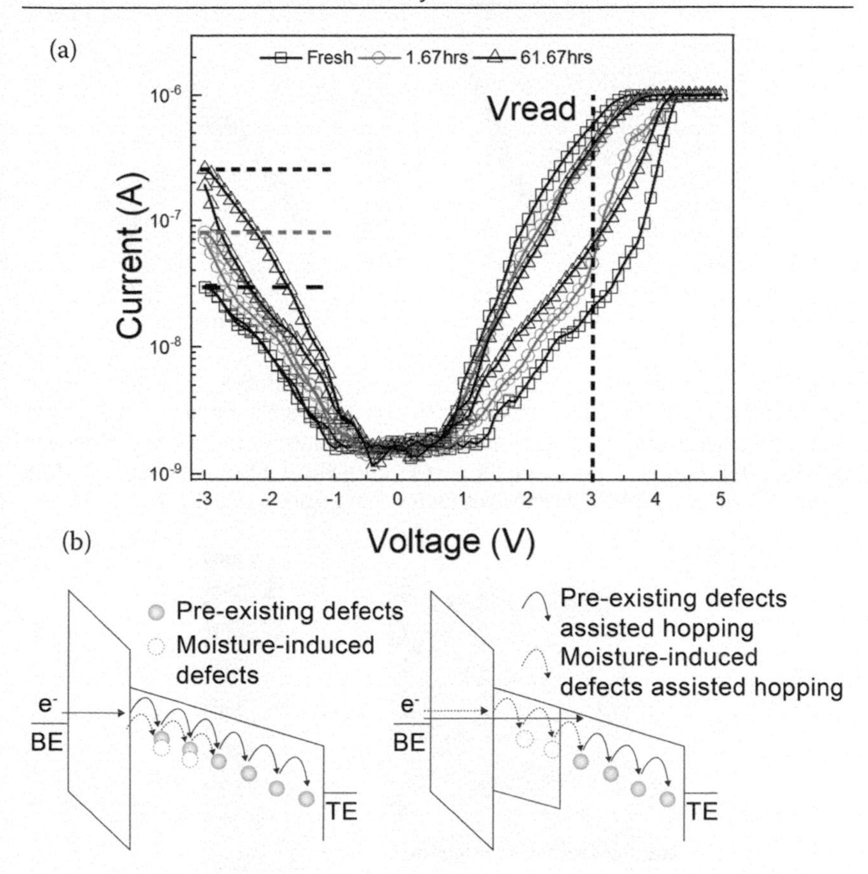

FIGURE 5.8 (a) I-V characteristics of VRRAM under 85°C/85%. (b) Energy band of LRS. (c) Energy band of HRS. © [2020] IEEE. Reprinted, with permission, from [Rui Gao, Effect of Moisture Stress on the Resistance of HfO_2/TaO_x-Based 8-Layer 3D Vertical Resistive Random Access Memory, IEEE Electron Device Letters, and Jan. 1, 2020].

layer-independent while μ_RHRS and σ_RHRS decline from the top layer to the bottom layer.[27,28] A successive length increase is observed from the high-resolution transmission electron microscope (HRTEM) image, as shown in Figure 5.10. Under the same SET/RESET voltage, the top layer devices suffer from a higher electric field compared to the lower layers, resulting in a fiercer sweep of traps and a wider defect-less region after reset operation, eventually giving rise to a wider barrier and higher resistance at HRS.

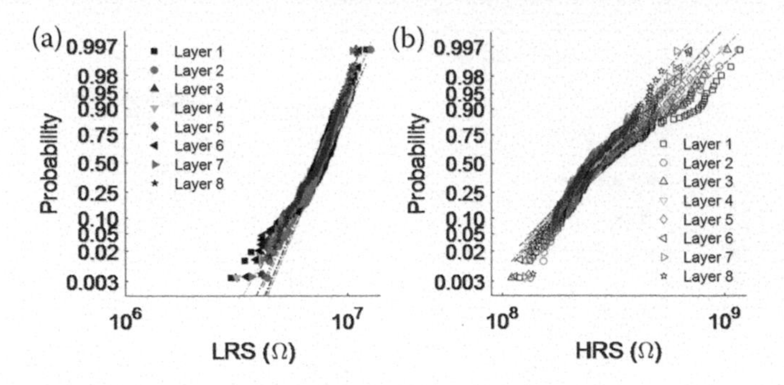

FIGURE 5.9 (a) The R_{LRS} distribution. (b) The R_{HRS} distribution. (c) The mean and variance of R_{HRS}. (d) The mean and variance of R_{HRS} across eight layers. 5.9(a, b) © [2019] John Wiley and Sons. Reprinted, with permission, from [Rui Gao, Dengyun Lei, Zhiyuan He, et al., Layer-dependent resistance variability assessment on 2048 8-layer 3D vertical RRAMs, Electronics Letters, and Aug. 1, 2019].

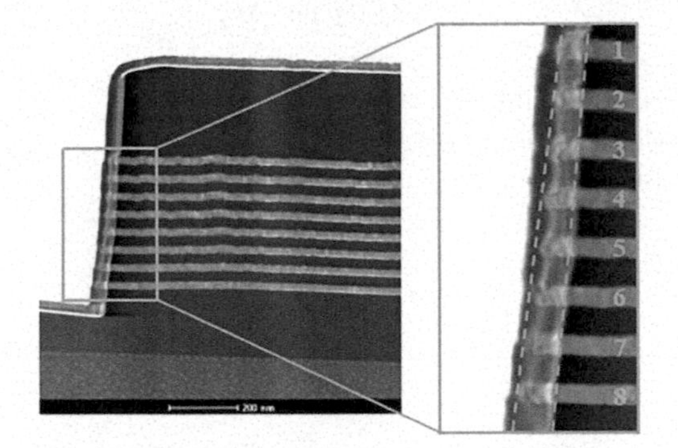

FIGURE 5.10 High-resolution transmission electron microscope (HRTEM) image shows a continuous length increase. © [2019] John Wiley and Sons. Reprinted, with permission, from [Rui Gao, Dengyun Lei, Zhiyuan He, et al., Layer-dependent resistance variability assessment on 2048 8-layer 3D vertical RRAMs, Electronics Letters, and Aug. 1, 2019].

REFERENCES, BIBLIOGRAPHY, OR WORKS CITED

[1] T. Gong, Q. Luo, H. Lv, X. Xu, J. Yu, P. Yuan, D. Dong, C. Chen, J. Yin, L. Tai, X. Zhu, S. Long, Q. Liu, and M. Liu, "Switching and failure mechanism of self-selective cell in 3D VRRAM by RTN-based defect tracking technique." In: *IEEE IMW* (May 2018).

[2] B. Gao, J. Kang, H. Zhang, B. Sun, B. Chen, L. Liu, X. Liu, R. Han, Y. Wang, Z. Fang, H. Yu, B. Yu, and D. Kwong, "Oxide-based RRAM: Physical based retention projection." In: *Proc. ESSDERC* (Sep. 2010), pp. 392–395.
[3] J. Lee, C. Du, K. Sun, E. Kioupakis, and W. Lu, "Tuning ionic transport in memristive devices by graphene with engineered nanopores," *ACS Nano*, vol. 10, no. 3, pp. 3571–3579, Mar. 2016.
[4] H. Tanaka, M. Kido, K. Yahashi, M. Oomura, R. Katsumata, M. Kito, Y. Fukuzumi, M. Sato, Y. Nagata, Y. Matsuoka, Y. Iwaka, H. Aochi, and A. Nitayama, "Bit cost scalable technology with punch and plug process for ultra high density flash memory." In: *VLSI Symp. Tech. Dig.* (Jun. 2007).
[5] Q. Luo, X. Xu, T. Gong, H. Lv, D. Dong, H. Ma, P. Yuan, J. Gao, J. Liu, Z. Yu, J. Li, S. Long, Q. Liu, and M. Liu, "8-layers 3D vertical RRAM with excellent scalability towards storage class memory applications," In: *IEDM Tech. Dig.* (Dec. 2017), pp. 48–51.
[6] E. Hsieh, P. Lu, S. Chung, K. Chang, C. Liu, J. Ke, C. Yang, and C. Tsai, "The experimental demonstration of the BTI-induced breakdown path in 28nm high-k metal gate technology CMOS devices." In: *VLSI Symp. Tech. Dig.* (Jun. 2014).
[7] Z. Chai, J. Ma, W. Zhang, B. Govoreanu, E. Simoen, J. Zhang, Z. Ji, R. Gao, G. Groeseneken, and M. Jurczak, "RTN-based defect tracking technique: Experimentally probing the spatial and energy profile of the critical filament region and its correlation with HfO_2 RRAM switching operation and failure mechanism." In: *VLSI Symp. Tech. Dig.* (Jun. 2016).
[8] J. Ma, Z. Chai, W. Zhang, B. Govoreanu, J. Zhang, Z. Ji, B. Benbakhti, G. Groeseneken, and M. Jurczak, "Identify the critical regions and switching/failure mechanisms in non-filamentary RRAM (a-VMCO) by RTN and CVS techniques for memory window improvement." In: *IEDM Tech. Dig.* (Dec. 2016), pp. 564–567.
[9] S. Lee, H. Cho, Y. Son, D. Lee, and H. Shin, "Characterization of oxide traps leading to RTN in high-k and metal gate MOSFETs." In: *IEDM Tech. Dig.* (Dec. 2009), pp. 763–766.
[10] J. Zou et al., "New insights into AC RTN in scaled high-κ/metal-gate MOSFETs under digital circuit operations." In: *VLSI Symp. Tech. Dig.* (2012), pp. 139–140.
[11] P. Huang, D. Zhu, C. Liu, Z. Zhou, Z. Dong, H. Jiang, W. Shen, L. Liu, X. Liu, and J. Kang, "RTN based oxygen vacancy probing method for Ox-RRAM reliability characterization and its application in tail bits." In: *IEDM Tech. Dig.* (Dec. 2017), pp. 525–528.
[12] F. Puglisi, L. Larcher, A. Padovani, and P. Pavan, "A complete statistical investigation of RTN in HfO_2-based RRAM in high resistive state." *IEEE Trans. Electron Devices*, vol. 62, no. 8, pp. 2606–2613, Aug. 2015.
[13] C. Chang, S. Chung, Y. Hsieh, L. Cheng, C. Tsai, G. Ma, S. Chien, and S. Sun, "The observation of trapping and detrapping effects in high-k gate dielectric MOSFETs by a new gate current Random Telegraph Noise (IG-RTN) approach." In: *IEDM Tech. Dig.* (Dec. 2008).

[14] J. Lee, J. Lee, J. Park, S. Chung, J. Roh, S. Hong, I Cho, H. Kwon, and J. Lee, "Extraction of trap location and energy from random telegraph noise in amorphous TiO_X resistance random access memories." *Appl. Phys. Lett.*, vol. 98, no. 14, p. 143502, Apr. 2011.

[15] R. Waser and M. Aono, "Nanoionics-based resistive switching memories," *Nature Mater*, vol. 6, no. 11, pp. 833–840, Nov. 2007.

[16] M. Zangeneh and A. Joshi, "Design and optimization of non-volatile multibit 1T1R resistive RAM," *IEEE Trans. Very Large Scale Integr. (VLSI) Syst.*, vol. 22, no. 8, pp. 1815–1828, Aug. 2014.

[17] F. Puglisi, and P. Pavan, "RTN analysis with FHMM as a tool for multi-trap characterization in HfOx RRAM." In: *EDSSC* (Jun. 2013).

[18] Z. Chai, J. Ma, W. Zhang, B. Govoreanu, E. Simoen, J. Zhang, Z. Ji, R. Gao, G. Groeseneken, and M. Jurczak, "RTN-based defect tracking technique: Experimentally probing the spatial and energy profile of the critical filament region and its correlation with HfO2 RRAM switching operation and failure mechanism." In: *VLSI Symp. Tech. Dig.* (Jun. 2016).

[19] J. Ma, Z. Chai, W. Zhang, B. Govoreanu, J. Zhang, Z. Ji, B Benbakhti, G. Groeseneken, and M. Jurczak, "Identify the critical regions and switching/ failure mechanisms in non-filamentary RRAM (a-VMCO) by RTN and CVS techniques for memory window improvement." In: *IEDM Tech. Dig.* (Dec. 2016). pp. 564–567.

[20] P. Huang, D. Zhu, C. Liu, Z. Zhou, Z. Dong, H. Jiang, W. Shen, L. Liu, X. Liu, and J. Kang, "RTN based oxygen vacancy probing method for Ox-RRAM reliability characterization and its application in tail bits." In: *IEDM Tech. Dig.* (Dec. 2017). pp. 525–528.

[21] N. Raghavan, R. Degraeve, L. Goux, A. Fantini, D.J. Wouters, G. Groeseneken, and M. Jurczak, "RTN insight to filamentary instability and disturb immunity in ultra-low power switching HfOx and AlOx RRAM." In: *VLSI Symp. Tech. Dig.* (Jun. 2013).

[22] N. Raghavan, R. Degraeve, A. Fantini, L. Goux, S. Strangio, B. Govoreanu, D.J. Wouters, G. Groeseneken, and M. Jurczak, "Microscopic origin of random telegraph noise fluctuations in aggressively scaled RRAM and its impact on read disturb variability." In: *IRPS* (Apr. 2013).

[23] C. Chang, S. Chung, Y. Hsieh, L. Cheng, C. Tsai, G. Ma, S. Chien, and S. Sun, "The observation of trapping and detrapping effects in high-k gate dielectric MOSFETs by a new gate current random telegraph noise (IG-RTN) approach." In: *IEDM Tech. Dig.* (Dec. 2008).

[24] H. Lv et al., "BEOL based RRAM with one extra-mask for low cost, highly reliable embedded application in 28 nm node and beyond." In: *IEDM Tech. Dig.* (Dec 2017), pp. 36–39.

[25] S. Choi, Y. Yang, and W. Lu, "Random telegraph noise and resistance switching analysis of oxide based resistive memory," *Nanoscale*, pp. 400–404, 2014.

[26] B. Oh, H. Cho, H. Kim, and H. Shin, "Observation of three-level random telegraph noise in GIDL current of Saddle-Fin type DRAM cell transistor." In: *IMW* (May 2010).

[27] N. Tega et al., "Anomalously large threshold voltage fluctuation by complex random telegraph signal in floating gate flash memory." In: *IEDM Tech. Dig.* (Dec. 2006), pp. 1–4.

[28] M. Uren, M. Kirton, and S. Collins, "Anomalous telegraph noise in small-area silicon metal-oxide-semiconductor field-effect transistors," *Phys. Rev. B*, vol. 37, no. 14–15, pp. 8346–8350, May 1988.

Applications of 3D RRAM beyond Storage

6

Xumeng Zhang, Xiaoxin Xu, and Jianguo Yang

Contents

6.1 NEUROMORPHIC COMPUTING

Neuromorphic computing technology, inspired by the human brain (Figure 6.1a), is expected to solve the "unsustainable development" dilemma of current AI platforms in computing power and energy consumption.[1]

DOI: 10.1201/9781003391586-6

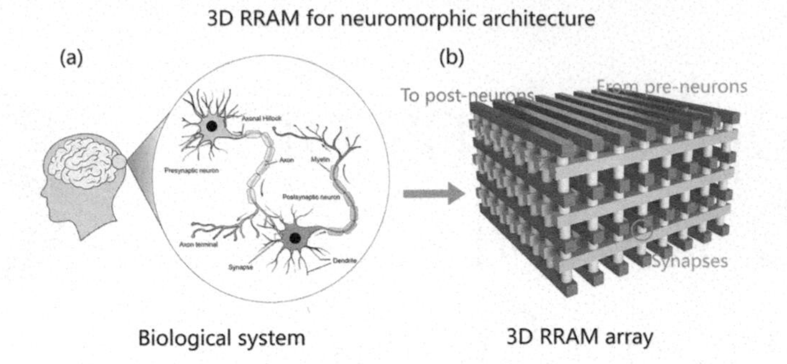

FIGURE 6.1 Schematic diagram of (a) human brain and the cortical neural systems vs (b) neuromorphic architecture based on a 3D integrated RRAM array.

Neuromorphic platforms have many novel characteristics compared to conventional computers,[2] such as in-memory computing, spike-based processing, event-based and asynchronous communication, and adaptive learning in hardware et al. Spiking neurons and plastic synapses are two fundamental units for constructing neuromorphic systems. Thus, it is very intriguing to build a neuromorphic computer through the hardware implementation of these two components, just like the human brain does. Resistive random-access memory (RRAM),[3] as one of the typical memristors, has proven advantages of high bionics, high integration density, low energy consumption, and so on, providing a new physical basis for the realisation of artificial neurons and synapses. The integrated RRAM array in a 3D structure, resembling the human brain architecture (Figure 6.1b) in particular, has great potential in becoming an artificially intelligent brain.

6.2 ARTIFICIAL SYNAPSES

In biological neural systems, the synapse is the most numerous component ($\sim 10^{15}$).[4] It is responsible for the complex connections between neurons and plays a vital role in the learning and memory activities of organisms. Currently, most works of memristor-based neuromorphic computing focus on emulating related functions of biological synapses. These functions could be divided into two types: long-term synaptic plasticity, and short-term synaptic plasticity.[5]

When using memristors to emulate the long-term plasticity of synapses, we need nonvolatile memristors, devices with good retention characteristics.

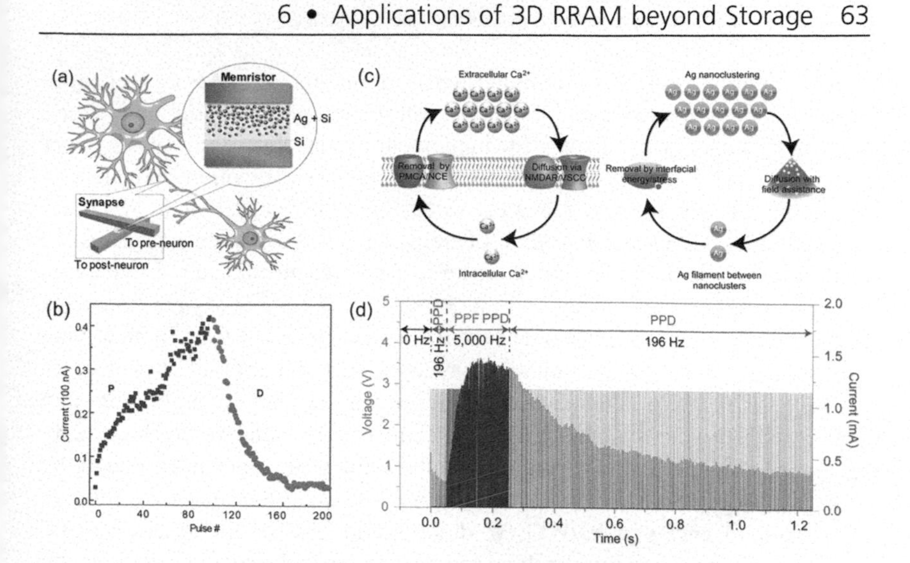

FIGURE 6.2 (a) Memristor for artificial synapses. (b) LTP and LTD properties implemented in memristor-based synapses. (c) Diagram of Ca^{2+} in biological synapse and the movement process of active ions in memristor. (d) Short-term plasticity implemented in an Au/SiOxNy:Ag/Au memristor.[6,8]

In addition, the nonvolatile memristors should better possess multilevel resistance states. In 2010, Lu et al. from the University of Michigan experimentally emulated the long-term plasticity of biological synapses based on a W/Si/Si: Ag/Cr/Pt memristor for the first time,[6] as shown in Figure 6.2a. Initially, the Si:Ag layer of the device is in a low resistance state and the Si layer is in a high resistance state. During the forward voltage scanning, the Ag atom will move to the Si layer, making the Ag interface gradually move down the electrode, and the conductivity of the device gradually increases. When the device is scanned by the reverse voltage, the Ag interface moves in the opposite direction, and the conductivity of the device decreases. In consequence, the intrinsic regulation of the conductance of the device can be used to emulate the plasticity of biological synapses. By applying continuous positive pulses to the device, the current flowing through the device gradually increases, corresponding to the long-term potentiation (LTP) behaviour of biological synapses. With the negative pulse applied to the device, the current flowing through the device gradually decreases, corresponding to the long-term depression (LTD) behaviour of the synapse, as shown in Figure 6.2b. This work shows that memristor can be used to emulate the long-term plasticity behaviour of biological synapses, opening the way for creating 2,000 artificial synapses.

In contrast with long-term plasticity emulation, the device generally needs volatile characteristics[7] to emulate short-term plasticity. In memristors, there are usually two schemes for implementing the short-term plasticity of synapses. One is realised by the oxidation-reduction reaction and electromigration process of active metal electrodes under the electric field; the other is realised by the electromigration of defects (such as oxygen vacancies) in materials. When emulating the short-term plasticity mechanism of synapses, the movement of active metal ions (such as Ag^+ or Cu^{2+}) or oxygen vacancies in the dielectric layer is regarded as the diffusion process of Ca^{2+} in biological synapses, so it is highly similar to synaptic ion dynamics,[8] as shown in Figure 6.2c. In 2011, T. Ohno et al. implemented the short-term plasticity in a $Ag/Ag_2S/Pt$ atomic switch under low-frequency stimulation by accurately regulating the quantum conductance.[9] In the same year, Lu et al. emulated the short-term plasticity mechanism of synapses by using the migration process of oxygen vacancies in $Pd/WO_x/W$ devices and realised the transformation from short-term memory to long-term memory of synaptic devices by repeatedly applying stimuli.[10] However, in these works, only the synaptic pair-pulse facilitation (PPT) behaviour has been realised, and the synaptic pair-pulse depression (PPD) behaviour in these devices has not been reported. Until 2016, Z. Wang et al. realised both the PPF and PPD synaptic behaviours,[8] by preparing an Au/SiO_xN_y:Ag/Au memristor by doping process, which linked the depletion process of Ag atoms with PPD, as shown in Figure 6.2d. So far, the work of realising synaptic short-term plasticity bionics by using the dynamic effect of ions has been verified in a large number of material systems, such as biopolymer films,[11,12] two-dimensional materials (h-BN),[13] organometallic compounds, metal oxides.[14] In these works, Y. Park et al. also implemented the PTP synaptic short-term effect based on an Au/lignin/ITO/PET memristor.[11] As reported, the short-term plasticity mechanism of biological synapses plays an important role in the computation process of biological neural systems, and the verification of computing function by using the short-term effect of artificial synapses needs to be further studied.

In biological systems, the cognition and learning processes usually follow some rules, which determine how the connected synapses perform long-term plasticity. Generally, the learning rules can be divided into spike-timing-dependent plasticity (STDP) and spike-rate-dependent plasticity (SRDP) in biology. Next, we will introduce the biological basis and device implementation of these two learning rules.

It is reported that LTP behaviour occurs when the synapse is under a high-frequency stimulation, which could remove the blocked Mg^{2+} on the NMDA receptor, while a low-frequency stimulation cannot.[15] However, if the presynaptic receives low-frequency stimulation and the postsynaptic receives stimulation signals as well, the membrane potential of the postsynaptic membrane will be polarised to remove Mg^{2+}. At this time, the postsynaptic will interact with the glutamate neurotransmitter released by the presynaptic to open the Ca^{2+} channel and enhance the strength of the synapse. This is the so-called Hebbian learning rule.[16] The study further found that the timing difference between the pre-neuron spikes and post-neuron spikes could lead to the change of synaptic weight in different directions.[17] When the timing of the pre-synaptic signal is earlier than that of the post-synaptic signal, it will cause long-term enhancement of synapse. On the contrary, the weight of the synapse will be weakened. Also, the change in synaptic weight is related to the spike timing difference before and after synapse. The smaller the timing difference is, the greater the change of weight is. According to such characteristics, researchers obtained the commonly used STDP learning rule.

Currently, there are mainly two methods to realise the STDP learning rules using memristors: one is to use overlapped pre- and post-pulses through a specified waveform design, and the other is to use nonoverlapped pulses in two-order memristors. One electrode of the device acts as a presynaptic membrane to receive stimulation signals from pre-neurons, and the other electrode acts as a postsynaptic membrane to receive action potential signals from post-neurons. For the overlapped scheme, the programmed pulse waveforms applied to the device usually have both positive and negative voltage values. The voltage value of a single polar pulse cannot change the resistance state of the device. According to the waveform design of the voltage, the difference in the effective voltage amplitude caused by the different intervals between the pre- and post-pulses leads to different adjustments to the device conductance, so as to obtain the device conductance modulation related to the timing difference between the pre- and post-pulses. However, the successful implementation scheme of nonoverlap does not require the superposition of pre- and post-pulses. Based on the above two programming schemes, researchers have carried out a lot of work in recent years. For example, S. Yu et al.,[18] Y. Li et al.,[19] and Prezioso et al.,[20] successively implemented STDP in different memristor devices in the mode of overlap. Figure 6.3a shows the pulse waveform used in Y. Li's work. The corresponding STDP curve is shown in Figure 6.3b.[19] In addition, Li et al. and Prezioso et al. created various forms of STDP curves in the device by designing special pulse waveforms, enhancing the possibility of the memristor in realising practical synaptic learning rules.

Compared with the form of overlap, the nonoverlap programming scheme has a higher bionic degree. In 2015, C. Du et al. used the dynamic characteristics in the second-order (or higher-order) memristor to implement STDP.[21] In this work, the nonoverlap pulse is used, as shown in Figure 6.3c, and the corresponding STDP curve is presented in Figure 6.3d. Except for using the memristor's high-order

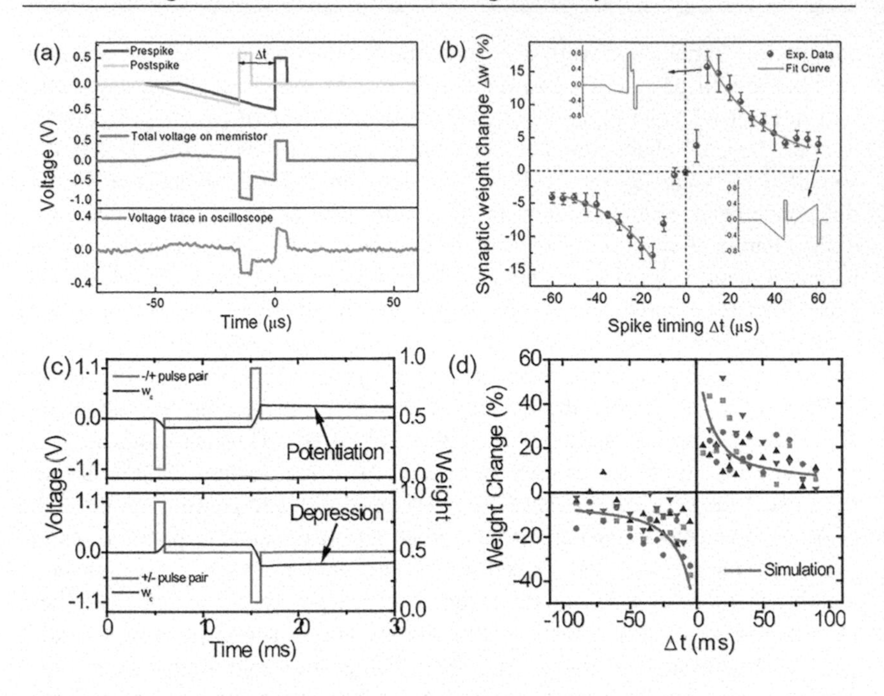

FIGURE 6.3 (a) Pulse waveform for implementing overlapped STDP. (b) The demonstrated STDP curve using the overlapped scheme. (c) Pulse waveform for implementing nonoverlapped STDP. (d) The demonstrated STDP curve using the nonoverlapped scheme.[19,21]

dynamics to implement STDP, Z. Wang et al. connected a volatile diffusive memristor and a nonvolatile drift memristor to implement STDP, in which the diffusive memristor is used as the timing control unit. Since the above STDP is determined by the timing difference between the pre- and post-neuron spike pairs, it is also called pair STDP. In addition to this form, there is also the form of triple STDP,[22] which has been reported in memristor-based synapses.[23,24]

SRDP is another commonly used learning rule of synaptic plasticity. It adjusts the synaptic weight according to the relationship between the frequency of pre- and post-neuron action potentials. BCM (Bienenstock, Cooper, and Munro) theory is a typical representative of SRDP learning rules.[25] BCM theory puts forward the threshold drift phenomenon of LTP or LTD and points out that the synaptic weight can be dynamically adjusted according to the average frequency of post-neuron action potential, which is history-dependent, as shown in Figure 6.4a.[26] According to the BCM model, when pre-neurons are excited if post-neurons fire high-frequency action potentials or the concentration of Ca^{2+} in post-synapse is relatively high, the synapse tends to achieve LTP. Or else, the synapse tends to have an LTD process.

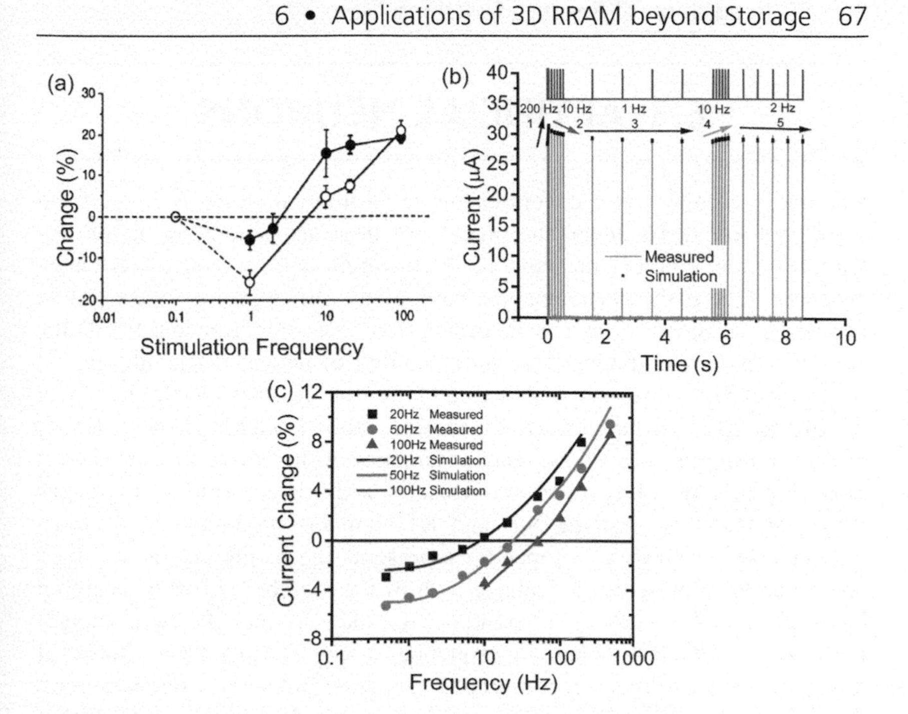

FIGURE 6.4 (a) SRDP learning rule in biological systems. (b) The history-dependent response of memristor-based synapses. (c) SRDP curve implemented in memristor-based synapses.[21,26]

In memristor, the SRDP learning rule is usually realised by applying pulses with different frequencies to the device and observing the dynamic response of the device under different frequency stimulation, as shown in Figure 6.4b.[21] It can be seen that when the device receives high-frequency stimulation (200 Hz) at the beginning, the response current increases, corresponding to the LTP process. When low-frequency stimulation (10 Hz or 1 Hz) is applied after high-frequency stimulation, the response current of the device decreases, corresponding to the LTD behaviour. However, after 1 Hz low-frequency stimulation, a 10-Hz low-frequency stimulation is applied to the device, the response current of the device will increase accordingly, and LTP behaviour will occur. Therefore, the response of synaptic devices is history-dependent. Figure 6.4c shows the effect of post-neuron frequencies on the conductivity change of the device under different historical stimulation frequencies. It can be seen that when the historical stimulus frequency is low, the turning frequency of LTD and LTP behaviour is low. With the historical stimulus frequency increasing, the turning frequency increases gradually. SRDP learning rules have been verified in several research groups' works. Recently, the BCM learning rule based on memristor has made the latest progress. Z. Wang et al. verified the BCM learning rule by using the $Pt/WO_{3-x}/W$ memristor and built a two-layer neural network simulation to realise the orientation selectivity.[24]

6.3 ARTIFICIAL NEURONS

Neuron is another critical component in biological systems to perform intelligent tasks. To unveil the process of neurons generating action potentials on the circuit or mathematically, different neuron models have been proposed. One is the phenomenological model, whose goal is to use simple mathematical abstractions (for example, the leaky integrate-and-fire [LIF] model[27]) to capture the input-output behaviour of neurons. The other is the biophysical model, whose goal is to emulate the electrophysiological state of the neuron membrane (for example, the Hodgkin-Huxley [H-H] model[28]). Figure 6.5a shows the schematic of the H-H neuron circuit model. In this model, two variable resistors with different turn-on voltages (R_{Na} and R_K) represent the Na^+ and K^+ channels of biological neurons, respectively. Meanwhile, a capacitor represents the membrane, and a fixed resistance R_L represents the leakage path of the membrane. After receiving input stimuli, the membrane potential lifts and activates the two voltage-gated ion channels to open in sequence when the membrane potential surpasses a threshold value, resulting in an action potential. Compared with the H-H model, the LIF neuron model is relatively simple, as shown in Figure 6.5b. In this model, a capacitor functions as a membrane to integrate the input signal, R_L serves as the leakage path, and the threshold switch (or variable resistance) serves as the ion channel.

Volatile threshold switching RRAM which can serve as the basis of dynamic threshold switching and emulate biological neurons' ion channels, has received wide attention due to its simple circuit designs. In 2013, Pickett et al. built a neuristor using two nanoscale $Pt/Nb_2O_5/Pt$ memristors according to the H-H neuron circuit model.[29] Figure 6.6a presents the

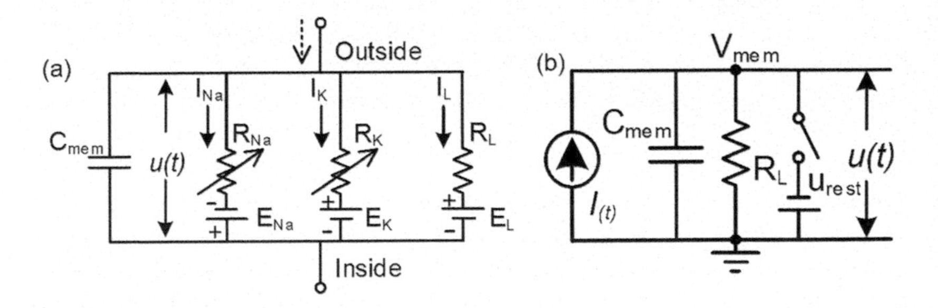

FIGURE 6.5 Schematic of (a) Hodgkin-Huxley (H-H) neuron circuit model and (b) leaky integrate-and-fire (LIF) neuron circuit model.

schematic of this neuristor. In this circuit, two NbO_2 memristors act as Na^+ ion channels and K^+ ion channels, respectively. Both channels consist of a memristor and a capacitance in parallel and are coupled by a load resistor. Figure 6.6b shows the typical I-V curve of this memristor. Such a neuristor can realise the neuronal behaviours of threshold firing, all-or-nothing action potential, lossless spike propagation, refractory period, tonic firing, and rapid burst firing. As shown in Figure 6.6c, no complete action potential generates when the input stimulus is relatively small (0.2 V), but a complete action potential happens when the input stimulus is large enough (0.3 V). As far as we know, this work is pioneering in the realisation of spiking neuron circuits using memristors, laying the foundation for scalable neuromorphic circuits.

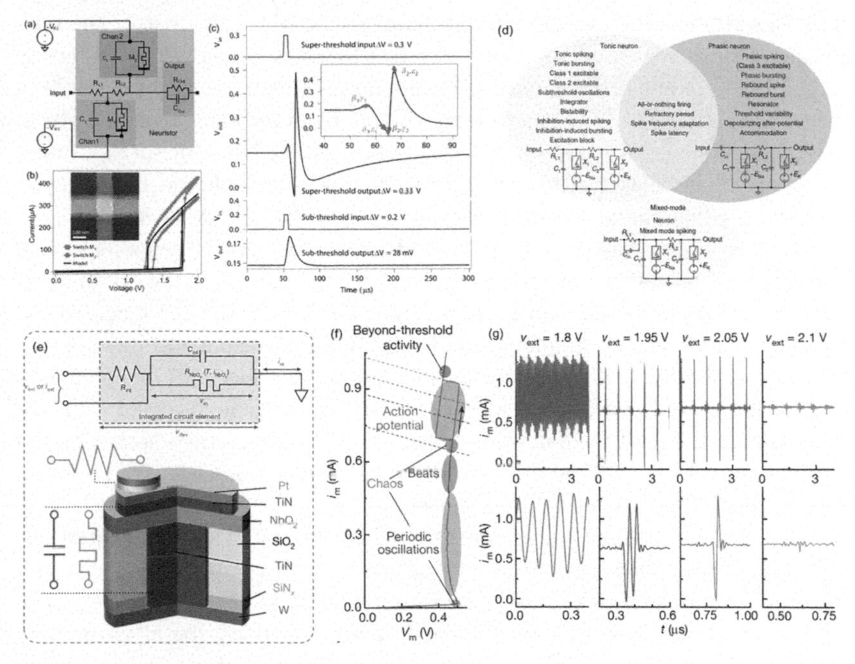

FIGURE 6.6 (a) Hodgkin-Huxley (H-H) neuron circuit implemented with NbO_2 memristors. (b) The typical I-V curve of NbO_2 memristors. (c) Spiking behaviour of the H-H neuron circuit. (d) Three prototype H-H neuron circuits and their demonstrated spiking behaviours. (e) Circuit model and the schematic illustration of the integrated nanocircuit element. (f) Quasistatic current–voltage behaviour of the circuit element. The dashed load-lines correspond to the colour-coded voltage biases shown in b. The overlaid ellipses indicate biasing regions that exhibit a range of qualitatively different oscillatory behaviours in the third-order element. (g) Measured temporal dynamics of the element's current (im) at different applied external voltages, as labelled.[29–31]

To create more better neuron firing modes, Yi et al. optimised the H-H neuron circuit model and built 23 biological neuron firing modes based on volatile neuron devices,[30] which more fully reflected the advanced nature of memristor-based neuron circuits. Figure 6.6d shows the three prototype neuron circuits and their demonstrated spiking behaviours. Different neuron firing modes are implemented by customising the passive R and C elements with no need for varying VO_2 device parameters. This work greatly simplifies the design and fabrication of a compact neuron circuit. In addition, Kumar et al. further achieved multiple neuron firing modes in a single integrated third-order neuristor by integrating the NbO_x memristor with a stacked external resistor.[31] Figure 6.6e shows the schematic of the integrated nanocircuit element, consisting of a NbO_2 memristor coupled with an internal parallel capacitor and an internal series resistor, as well as the structure of the NbO_2 memristor. The self-sustained sinusoidal oscillation occurs when the bias is below the hysteresis ($v_{ext} = 1.8$ V), while periodic two-spike bursting occurs when the bias is within the hysteresis ($v_{ext} = 1.95$ V), as shown in Figure 6f and 6g. This is resulted from that the neuron circuit works on different regions of the NbO_2 memristor under different voltage biases. This work provides a feasible solution for realising high-density neuron circuits to construct an efficient brain-like system.

Neuron circuits based on the H-H model have high requirements for device uniformity and parameter matching between circuits, increasing the difficulty of large-scale integration and application. Thus, the LIF model has received extensive attention in system integration for its simple structure and low computational complexity. In the LIF neuron circuit, only one capacitor is coupled with a volatile threshold switching memristor (TSM). The capacitor is responsible for integration while the TSM determines the threshold and generates spikes. Based on this principle, the LIF neuron circuits have been demonstrated and built on various TSMs.[32–35] Zhang et al. demonstrated a LIF neuron based on the Ag/SiO_2/Au TSM.[33] Figure 6.7a presents the schematic of the spiking neuron circuit. Figure 6.7b shows the firing characteristic of this artificial neuron, with increasing input pulse amplitude, the firing frequency increases obviously. The neuron circuit achieved four fundamental neuron functions: the all-or-nothing spiking of an action potential, threshold-driven spiking, a refractory period, and strength-modulated frequency response. In addition, the integration and leaky process could also be illustrated by the memristor's intrinsic dynamics. Wang et al. created a LIF neuron with stochastic dynamics based on a diffusive Pt/SiO_xN_y:Ag/Pt memristor (Figure 6.7c, d), where the migration of silver is similar to neurons' ion channels.[36] To enhance the functionality of such a neuron circuit, Zhang et al. further constructed a hybrid memristor-CMOS neuron, which has the basic LIF neuron function and enables the in-situ tuning of the connected synapses.[37] This work proposes a novel way to realise in situ learning for future neuromorphic computing systems.

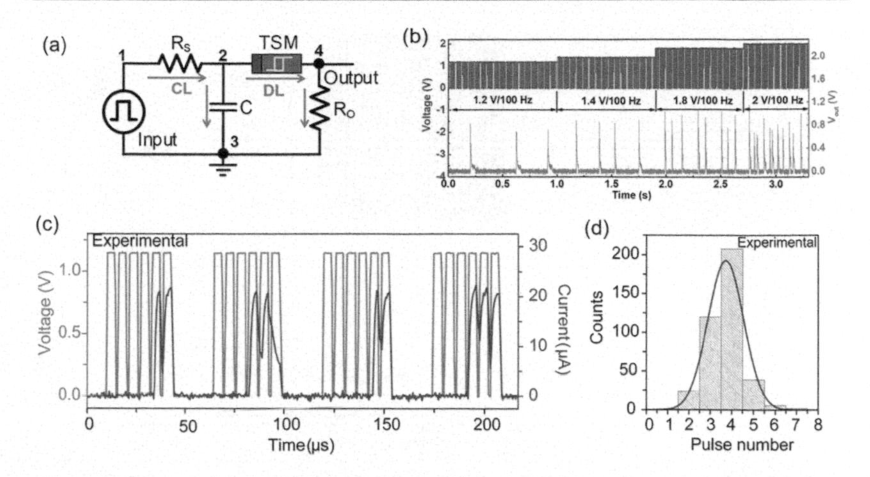

FIGURE 6.7 (a) LIF neuron circuit based on TSM. (b) Spiking behaviour of the TSM-based LIF neurons. (c) Leaky integrate-and-fire behaviour of the diffusive memristor. (d) Statistical results of the pulse number for firing, indicating a stochastic process.[33,36]

6.4 SNN BASED ON RRAM ARRAY

Considering the potential advantages of memristor-based synapses and neurons, researchers have carried out great efforts to build a neuromorphic system based on the spiking neural network (SNN) platforms. According to the types of used neuron circuits, the related work can be divided into CMOS neuron-assisted SNN and fully memristive SNN. In 2018, Prezioso et al. demonstrated a representative SNN work with memristor-based synapses and CMOS-based neurons.[38] In this work, M. Prezioso et al. used a 20 × 20 memristor synaptic array and CMOS neurons to verify STDP learning rules and realise correlation detection of input signals (Figure 6.8a). Figure 6.8b shows the schematic of related hardware. The neuron circuit used obeys the principle of the LIF neuron circuit. The first transconductance amplifier (TIA) is used to read the synaptic current from the memristor array, and the integrating amplifier integrates the read current. When the output voltage of the integration amplifier surpasses the set threshold value (V_{b2}), the last comparator outputs a high-level signal to trigger the generator to generate a voltage pulse with positive and negative polarity (Figure 6.8a), which can support the overlapped STDP learning rule in the memristor synapse. In the working process, the switch in the feedback loop of the pulse generator and the discharge loop of the integration amplifier are operated synchronously,

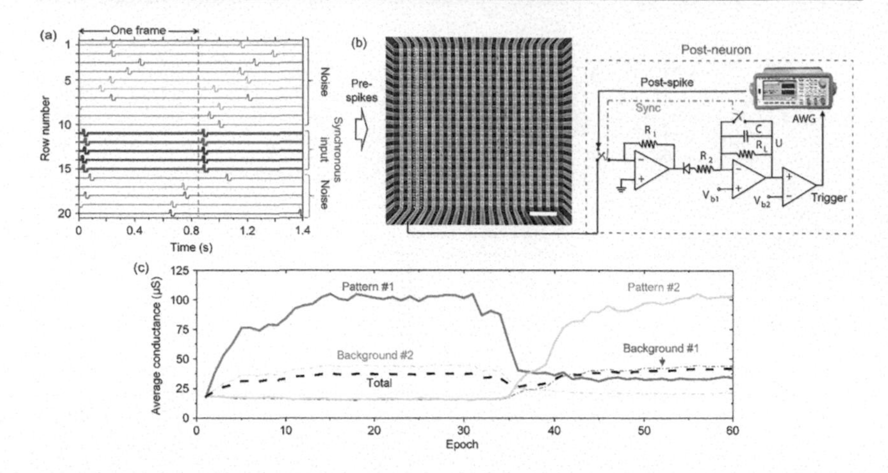

FIGURE 6.8 (a) Input signal patterns for coincidence detection. (b) SEM image of memristor synaptic array and schematic diagram of CMOS neuron. (c) Evolution of synaptic conductance versus iteration number under two input patterns.[38]

corresponding to different operations of the reasoning process and the training process, respectively. In the reasoning process, the switch in the feedback loop of the pulse generator and the discharge loop of the integration amplifier is disconnected at the same time; When the pulse generator triggers the action potential signal, both switches are closed at the same time. The connection of the feedback loop is used to transmit the post neuron action potential, while the connection of the discharge loop is used to discharge the capacitor. The correlation detection of the two low noise modes is further verified. Figure 6.8c shows the evolution process of memristor synaptic conductance under applying two different correlation patterns. In addition, D. Ielmini et al. have also studied some simulation work based on memristor synapse array and CMOS neuron circuit,[39–41] which preliminarily confirmed the feasibility of constructing a SNN chip by using memristor synapse array.

Due to the complex structure of CMOS neuron circuits, which is not conducive to large-scale integration, the use of memristor neurons to build neuromorphic systems has gradually become a research hotspot. In 2018, Wang et al. used memristor neurons and synapses to build an 8 × 8 fully memristive spiking neural network for the first time in the world.[36] The schematic diagram of its hardware structure is shown in Figure 6.9a, b. They monolithically integrated the network on chip, as shown in Figure 6.9c. Based on this system, the convolutional inference process is validated and unsupervised learning of input patterns is demonstrated. In addition, Duan et al. performed similar work through integrating the NbO_x-based neuron and $Ta/TaO_x/Pt$-based synapses.[42] To take full use of SNN in power consumption,

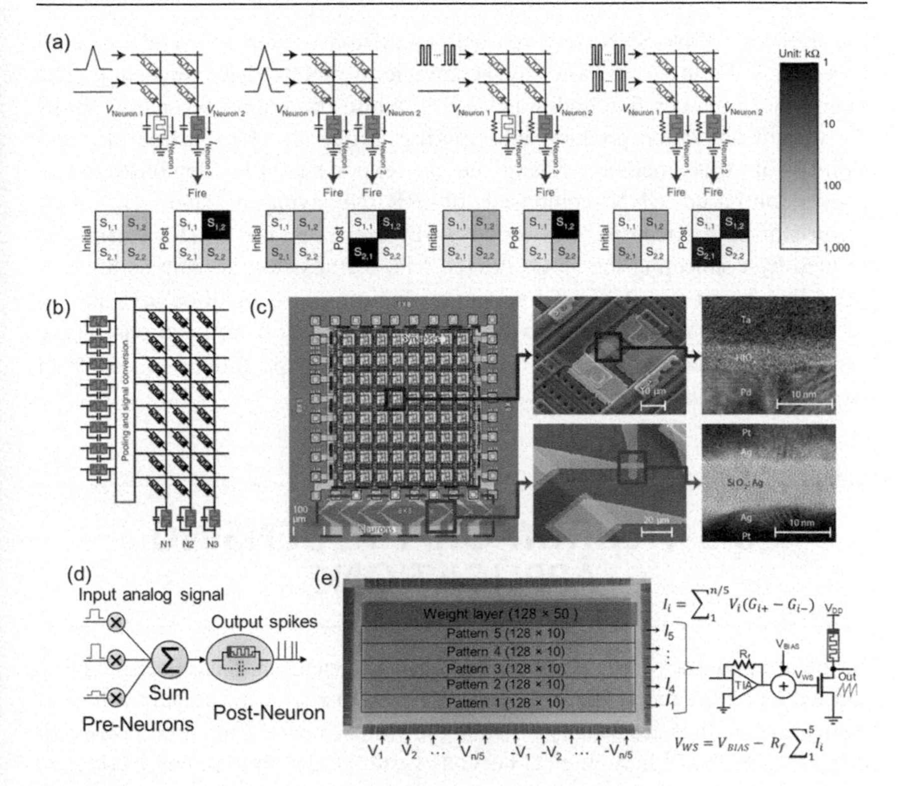

FIGURE 6.9 (a) The working principle of a fully memristive SNN. (b) The schematic of a 8 × 3 fully memristive network. (c) Monolithically integrated memristive network on chip. (d) Schematic of conversion based SNN. (e) A conversion-based SNN composed of RRAM synapses and 1T1R neurons.[36,44]

Zhang et al. further demonstrated a fully memristive temporal coding (TC) SNN, composed of five NbO_x memristor-based neurons and a 64 × 64 1T1R TaO_x/HfO_x memristor-based synaptic array.[43] Compared to the general rate coding (RC) SNN, this TC SNN shows a significant advantage in inference speed, power, and neuron's lifetime.

Since most of the commonly used data sets are for artificial neural networks (ANNs), the analogue quantity in the data set needs to be converted into pulse trains of corresponding frequencies when used for SNNs,[44] resulting in a loss of accuracy. The predicament of applying SNNs to process data needs to alleviated. To this end, it is an effective method to convert ANNs into SNNs by adjusting the weights and neuron parameters. The conversion method-based SNNs can have both the high energy efficiency of SNNs and the high precision of ANNs.[45,46] Based on this, Midya et al.

illustrated an ANN–SNN converter using a diffusive memristor and a parallel capacitor.[47] Figure 6.9d shows the schematic of ANN–SNN conversion. The input layer operates like traditional ANN, while the output layer transforms the weight activation products into spiking frequency. To illustrate the feasibility of this method, Zhang et al. experimentally demonstrated a conversion-based SNN composed of RRAM synapses and ten 1T1R memristor-based neurons,[44] as shown in Figure 6.9e. The 1T1R neuron was formed by connecting the NbO_x device with a transistor, serving as the rectified linear unit (ReLU) in the network, obtaining a recognition accuracy of up to 85.7% on MNIST, close to the results of ReLU software neurons. These results show that the 1T1R neurons are promising primitives to construct large-scale SNNs in the future.

6.5 NEUROMORPHIC SENSING APPLICATIONS

As mentioned above, memristor-based artificial neurons have been actively studied and explored to build efficient SNNs. However, the signals collected from surroundings are usually in analogue forms, which cannot be processed directly in SNNs.[48] In biological nervous systems, the afferent nerve converts the signals received from sensors into spikes and transmits them to central nervous systems for further processing. Therefore, to build an intelligent processing system that integrates sensing, storage, and calculation, it is necessary to construct a special unit to mimic the afferent nerve in biological systems (Figure 6.10a).

To meet the above requirement, Zhang et al. constructed a highly compact artificial spiking afferent nerve (ASAN) based on a 3D NbO_x memristor for the first time.[49] The most important component in the ASAN is the NbO_x oscillator composed of the NbOx memristor and a resistor, whose output frequency presents a quasi-linear relationship with the input voltage under normal stimuli and decreases under excessively strong stimuli. Furthermore, an ASAN-based artificial spiking mechanoreceptor system (ASMS) connected with a piezoelectric device was constructed, featuring no need for an external power source, as shown in Figure 6.10b. Figure 6.10c shows the frequency-voltage curve of the ASAN. In addition, the ASAN can be readily extended to process sensory signals from other sensors, such as smell, taste, sight, hearing, temperature, magnetic field, and humidity. In biological systems, signal sensing, generally in a multimode-fused form, helps people

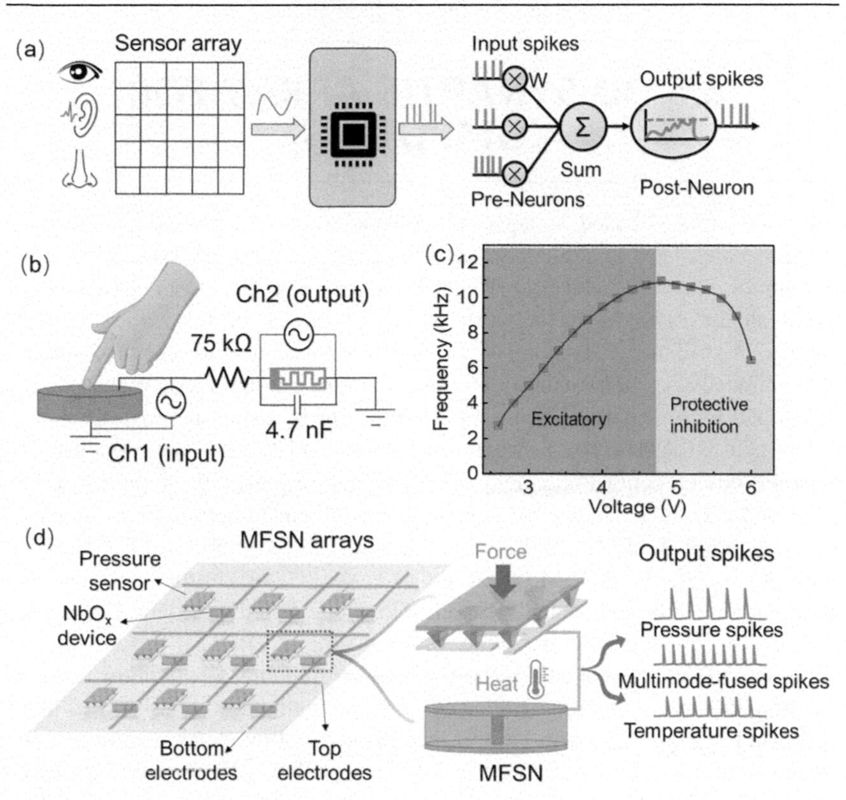

FIGURE 6.10 (a) Schematic diagram of neuromorphic sensing system. (b) Artificial spiking mechanoreceptor system. (c) Response curve of the afference nerve. (d) Schematic of the multimode-fused spiking neuron array based on memristors and pressure sensors.[49,50]

obtain comprehensive object properties and make accurate judgements. To emulate this function, Zhu et al. reported a multimode-fused spiking neuron (MFSN) array with a compact structure to achieve human-like multisensory perception,[50] as shown in Figure 6.10d. The MFSN heterogeneously integrates a pressure sensor to process pressure and a NbO_x-based memristor to sense temperature. Combining such an array with SNN, the authors demonstrated enhanced tactile pattern recognition, and successfully classified objects with different shapes, temperatures, and weights, validating the feasibility of our MFSNs for practical applications. These proof-of-concept MFSNs enable the building of multimodal sensory systems and contribute to the development of highly intelligent robotics.

6.6 3D SYNAPTIC ARRAY FOR COMPUTING

Although RRAM-based synapse is considered as a promising electronic synapse candidate. In human brains, the synaptic network consisting with billions of neurons is intricate 3D architecture. The present investigation is based on the 2D synaptic network which decreases the integration density greatly. High-density 3D synaptic networks are essential to realise innumerable connections among neurons in the brain. For the first time, Wang et al.[51] developed high-density 3D synaptic architecture by using the cost-efficiency 3D vertical RRAM with the structure of $Ta/TaO_x/TiO_2/Ti$. This proposed 3D synaptic device could implement several synaptic plasticity, including LTP, LDP and STDP. It shows the ultra-low power consumption (7.5 fJ/spike), which is comparable to the biological synapses. To improving the training accuracy of this 3D RRAM-based synapse, the authors further optimise the linearity of weight update and the fault tolerance of hardware neural networks (HNNs) by utilising a bipolar-pulse-training scheme (BPTS).[52] By using the additional offset pulse in BPTS, the noticeable rapid change of conductance at the first P/D-pulse is eliminated, thus achieving the low nonlinearity. The optimised synaptic characteristics are utilised to simulate the training evolution of an 8×8 binary alphabetic "B" pattern. The BP-scheme with lower nonlinearity showed a significant improvement in final training accuracy. The low-power, high-density, and reliable analogue 3D electronic synaptic array is significant for implementation of highly anticipated HNNs.

For wearable neuromorphic computing applications, it is important to investigate the mechanical flexibility of the artificial 3D synaptic network. Recently, flexible crystalline materials-based memristors, including 2D materials, metal-organic frameworks, covalent organic frameworks, and perovskites have been proposed for in-memory computing and artificial neural networks.[53–57] Wang et al.[57] designed a 3D flexible crossbar array based on ternary oxide (HfAlOx) via a low-temperature ALD technique at 130 °C. Grounded on the multilevel resistive switching characteristics of HfAlOx-based artificial synapse, the typical synaptic plasticity of a neural network can be mimicked by these memory units in each layer, including STP and LTP. The authors further demonstrated the mechanical flexibility of a 3D flexible network under a bending state (radius = 10 mm). The flexible HfAlOx-based memristor exhibits stable LTP/LTD behaviours under the bending state in three layers of the 3D network. Besides, the flexible memristors show stable resistive switching characteristics after folded 1,000 cycles. The development

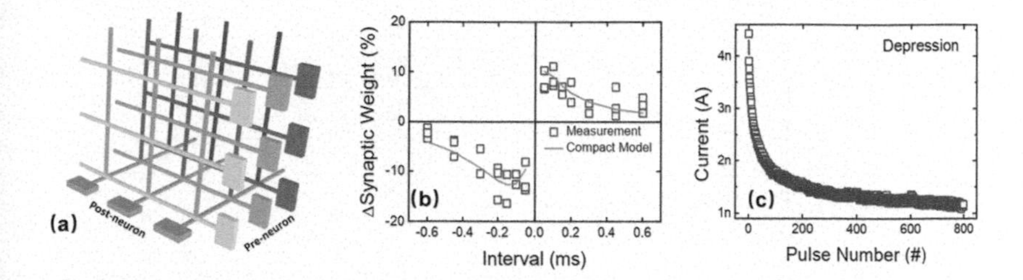

FIGURE 6.11 3D synaptic array with the structure of Ta/TaO_x/TiO_2/Ti for computing. (a) High-density 3D synaptic network emulates that intricate 3D synaptic network connects billions of neurons in human brains. (b) STDP measurement result in the 3D synaptic device. (c) Depressing characteristics in the 3D synaptic advice. © [2014] IEEE. Reprinted, with permission, from [I-Ting Wang, 3D synaptic architecture with ultralow sub-10 fJ energy per spike for neuromorphic computation, IEEE Proceedings, and Dec. 1, 2014].

of 3D flexible neuromorphic devices opens up many possibilities for the realisation of next-generation computing platforms with high comfort and conformability to biological systems (Figure 6.11).

6.7 NEUROMORPHIC HARDWARE DESIGN BASED ON THE 3D VRRAM

The simple two-terminal structure of RRAM naturally forms the two-dimensional (2D) crossbar array that enables large-scale parallel computations. Meanwhile, the 3D architecture of the memristor is a denser structure to match the ever-growing size of the neural network model (Figure 6.12a). Among the two common types of 3D structures, the 3D VRRAM with even/odd word line (WL) structure owns outstanding area efficiency and high potential in neural network implementation since this architecture contains doubled cell bits than the WL plane structure.

However, in the 3D VRRAM with an even/odd WL structure, a sophisticated control mechanism to access the cells and exacerbate the sneak path issue is required as the result of the increased interconnect lines. To minimise the side effects induced by the increase of interconnects, Kim et al.[58] proposed the hardware design through a more balanced structure along the three orthogonal directions. As shown in Figure 6.12b, the input data is fed through the vertical pillar electrodes and the output is detected at the WLs.

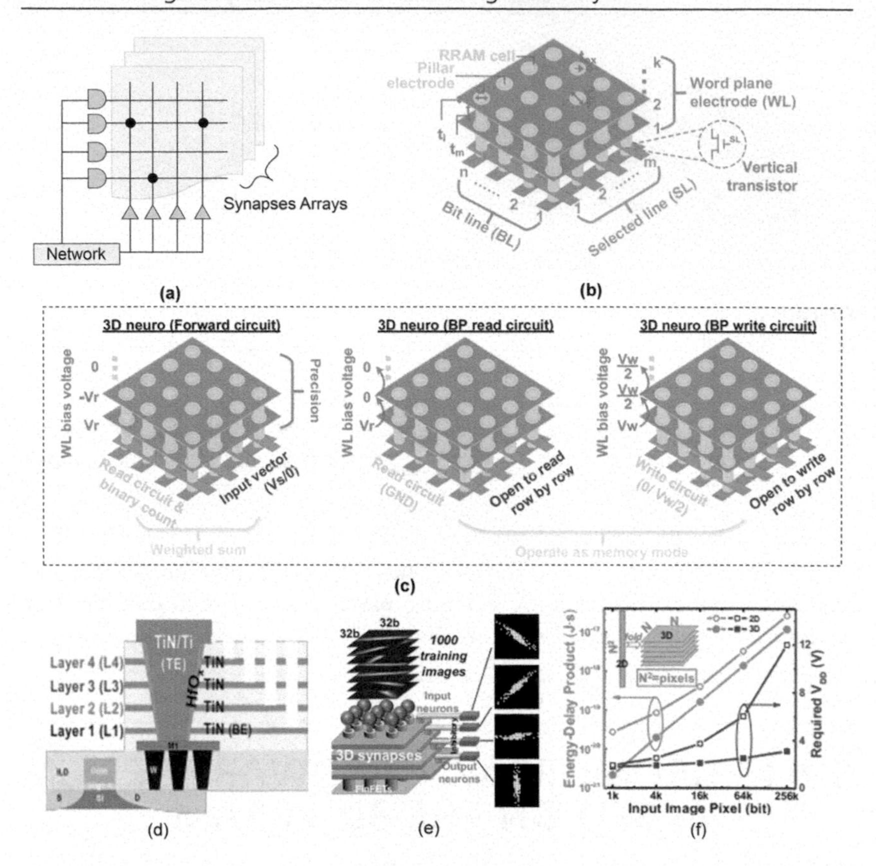

FIGURE 6.12 (a) 3D Neuromorphic IC Architecture. (b) Schematic of 3D V-RRAM architecture. (c) The operation mode of 3D V-RRAM array for neuromorphic computing. (d) Schematic of four-layer 3D RRAM integrated in FinFET platform. (e) A 3D neuromorphic visual system based on a 32 × 32 × 4 3D array. (f) SPICE-simulated energy-delay product and required VDD to programme worst-case–located synapses in 2D and 3D WTA networks.[58,60] 6.12(b–c) © [2017] IEEE. Reprinted, with permission, from [Zhiwei Li, Design of Ternary Neural Network With 3D Vertical RRAM Array, IEEE Transactions on Electron Devices, and Jun. 1, 2017]. 6.12(d–f) © [2016] IEEE. Reprinted, with permission, from [Haitong Li, Four-layer 3D vertical RRAM integrated with FinFET as a versatile computing unit for brain-inspired cognitive information processing, IEEE Proceedings, and Jun. 1, 2016].

This design provides a more balanced array configuration on layers and reduces the elongated sneak path efficiently. Furthermore, they define a weight adjustment method to harmonise with the use of even/odd lines and devise a sequential operation to ensure that only intended cells are programmed. The SSC with high nonlinearity is also utilised to mitigate the sneak path issue. This work verifies the effectiveness of the even/odd WL structure for neuromorphic system implementation.

For the neuromorphic hardware design based on the 3D VRRAM, the pillar electrodes are utilised as input vector and the plane electrodes as weighted-sum outputs.[59–62] Such operation scheme limits the number of output neurons that equals the number of vertical layers. Li et al.[58] proposed a novel operation scheme to combine the selected lines and the word-plane electrodes as input vector, and all the BLs are designed as weighted-sum outputs. Figure 6.12c demonstrates the operation scheme of 3D V-RRAM toward feedforward/backward inference (read) mode and weight update (write) mode. Compared to the 2D implementation, the proposed 3D V-RRAM implementation shows a larger write margin for weighted sum/ weight update, smaller latency, and energy consumption for weight update. This work demonstrates the attractiveness of building a monolithic 3D neuromorphic hardware platform. Li[60] improves the energy-delay product of neuromorphic computing by building the 3D vertical RRAM integrated with FinFET. Figure 6.12d exhibits the structure diagram. The 3D vertical RRAM is used for the system-level simulations, as shown in Figure 6.12e. In the unsupervised winner-take-all visual system, the 3D architecture "folds" neurons/synapses into the balanced plane with dense connections. The interconnect RC that effects and avoids long sneak leakage paths are thus reduced. As shown in Figure 6.12f, the 3D architecture improves the energy-delay product by 55% and reduces V_{DD} by 74%, which is a big improvement over the conventional 2D architecture.

Although the 3D VRRAMs are considered as the most promising candidate to increase the density of neuromorphic circuits, the complicated interconnection of 3D VRRAM makes the design more challenging, mainly due to several issues. For instance, these vulnerable unselected cells that share the same lines with the selected cells are more likely to be affected and the sneak leakage path per area increases. So, 3D VRRAM requires more carefully designed operations before building a monolithic 3D neuromorphic hardware platform.

6.8 COMPUTING-IN-MEMORY MACRO BASED ON 3D RRAM

The non-volatile computing-in-memory (nvCIM) schemes based on 2D RRAM have been widely investigated.[63–66] However, there is no nvCIM scheme dedicated to 3D vertical RRAM, which provides higher parallelism, capacity, and density for multiply-and-accumulate (MAC) operations. The 3D convolutional kernels, achieved by convolving a 3D kernel into a cube formed by stacking multiple pieces together, are the most important operations to achieve highly accurate image recognition by feature extraction. Huo et al.[67] implemented the 3D convolutional operations by the HfO_2/TaO_x-based eight-layer 3D VRRAM. Figure 6.13a is the schematic of a 3D convolution kernel

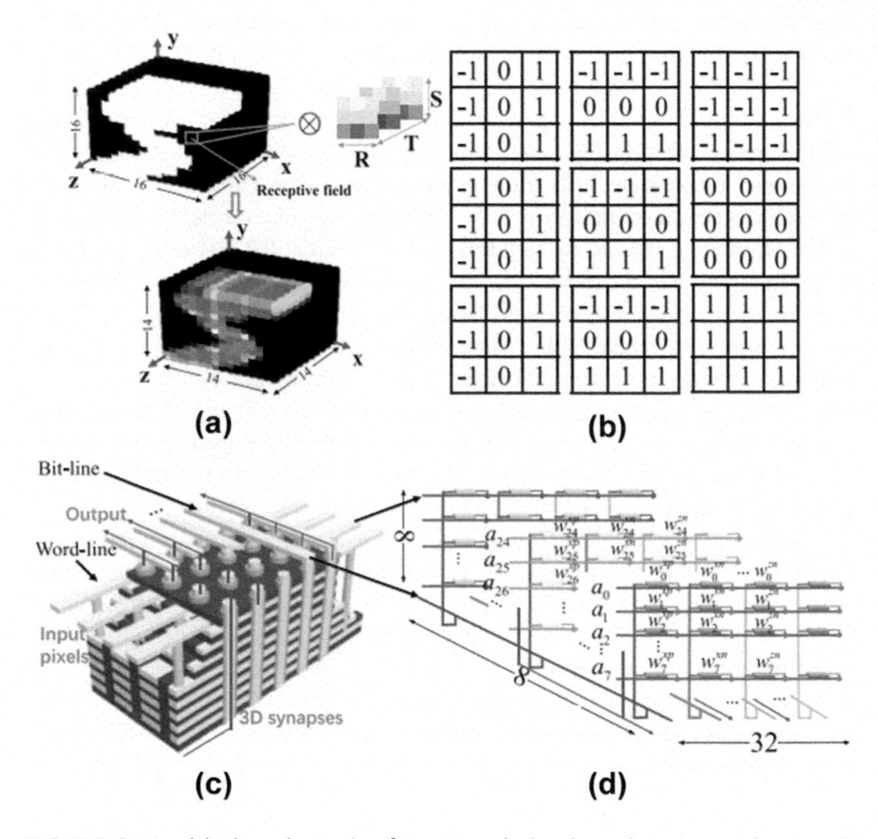

FIGURE 6.13 (a) The schematic of 3D convolution kernel on 3D version MNIST handwritten digits with 16 × 16 × 16 pixels. (b) 3D Prewitt kernel for detecting edge surfaces parallel to the y-z plane, x-z plane, and x-y plane, respectively. (c) The schematic of the 3D VRRAM architecture and current flow for one convolution operation. (d) The implementation of 3D Prewitt kernel Gx, Gy, and Gz on 3D VRRAM.[58,60] © [2020] IEEE. Reprinted, with permission, from [Qiang Huo, Demonstration of 3D Convolution Kernel Function Based on Eight-Layer 3D Vertical Resistive Random Access Memory, IEEE Proceedings, and Mar. 1, 2020].

on a 3D version MNIST handwritten digits with 16 × 16 × 16 pixels for demonstration. They use the Prewitt kernel to extract the 2D edge surface of a 3D image and display 3D convolution on 3D VRRAM. As shown in Figure 6.13b, the G_x, G_y, and G_z are 3D kernels used to detect the 2D edge surface parallel to the y-z plane, x-z plane, and x-y plane, respectively. The weights of Prewitt kernel are mapped to the binary 3D VRRAM architected as in Figure 6.13c. After convolving the 3D Prewitt kernels Gx, Gy, and Gz with the binary 3D digital input, the image pixel value of the target value is output. Figure 6.13d is the implementation schematic of 3D Prewitt kernel on 3D VRRAM. By utilising an FPGA-controlled relay-matrix based test platform, the 3D version of MNIST handwritten digit edge detection is well implemented on an eight-layer 3D VRRAM based on HfO_2/TaO_X. They verified that CIM implemented with 3D VRRAM arrays can correctly realise 3D convolutional kernels, and has higher parallelism, lower power consumption and higher capacity than traditional architectures.

The implementation of 3D CNN is facing other challenges, including more array area due to (a) the increase in the number of weights of CNN, (b) the need for multibit weight (W) of CNN for high recognition accuracy, and (c) the wider metal wire for huge bit line (BL) current (IBL) of conventional parallel WL input (IN) in situ MAC (PWIMAC) scheme. Figure 6.14b is data flow and architecture of the conventional parallel WL input in situ vector–matrix multiplication (PWIVMM) scheme for 1bIN–1bPW vector–matrix multiplication operation (white lines) and the ADINWM scheme for 8bIN–8bPW VMM operation. To overcome these challenges, Huo et al.[68] implements a CIM macro by 3D VRRAM. As shown in Figure 6.14a, the multilevel self-selective (MLSS) 3D VRRAM is combined with anti-drift multibit analogue IN-W multiplication (ADINWM) featuring low IBL to contrast a high-density computing scheme. Due to the drift of RRAM cell for the multilevel and high-capacity RRAM, IBL overlap of MAC computing cannot be effectively distinguished. For this issue, a current-amplitude-discrete-shaping (CADS) scheme is proposed to enlarge the sensing margin (SM) and eliminate the unrecoverable distortion of error accumulation.

Increasing the resistance of RRAM cell can reduce total output (OUT) IBL, and the total power due to the high-precision MAC operations. However, the small IBL would result in long latency (TAC). The 3D VRRAM used by Huo et al. shows the nA operation current, which could reduce the system power. So, in the 3D RRAM based CIM, an analogue multiplier (AM) combined with a gate precharge switch follower (GPPSF) and a direct small current converter (DSCC) are designed to reduce TAC. A 55-nm 2Kb 3D VRRAM-CIM macro presents the first 8bIN-9bW nvCIM macro, featuring the highest output precision, the highest energy efficiency (EFMAC) (135.59-8.41TOPS/W) and the best bit density (58.2 b/um^2) for MAC operations from binary to 8bIN-9bW-22bOUT

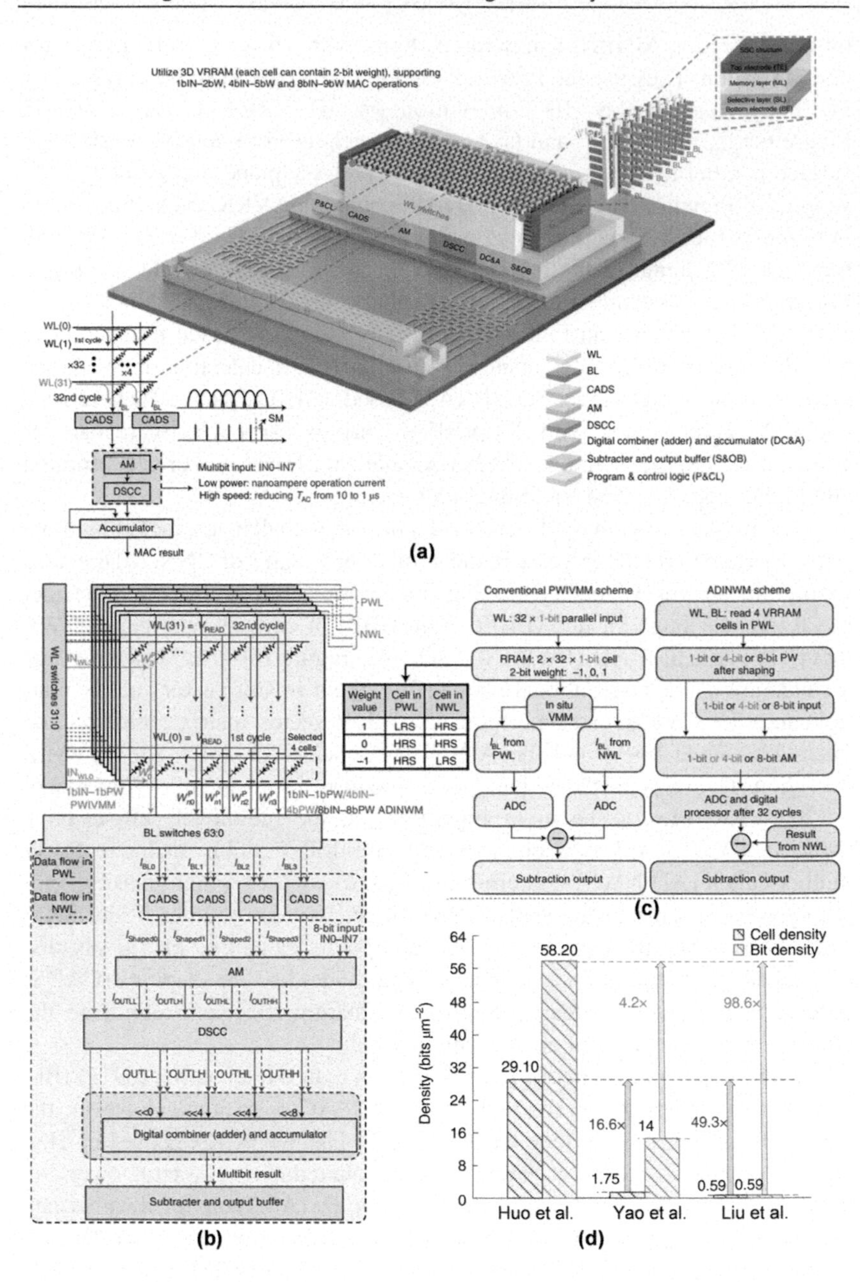

Weight value	Cell in PWL	Cell in NWL
1	LRS	HRS
0	HRS	HRS
−1	HRS	LRS

(caption on next page)

compute precisions. Figure 6.14c is the implementation framework of the conventional PWIVMM scheme and proposed ADINWM scheme. The macro can be configured as the 1bIN–2bW, 4bIN–5bW and 8bIN–9bW operations based on the ADINWM scheme. The power of our chip is at the nW level thanks to the higher resistance of 3D VRRAM cell, which is suitable for extremely low-power edge computation. The main power is produced by 3D VRRAM, while the DSCC consumes the most energy for 4bIN-5bW and 8bIN-9bW operations. Besides, the number of cells per unit area of our VRRAM is far higher than that of 2D RRAM, and the bit density is more than ten times that of in the past. Figure 6.14d demonstrates the bit density of this micro. This scheme for AI edge computation is also potentially useful in the design of nvCIM macros based on other emerging 3D RRAM technologies that may suffer from device stability issues.

6.9 3D RESERVOIR COMPUTING BASED ON 3D DYNAMIC MEMRISTORS

Driven by the rapid development of the Internet of Things, the amount of temporal sensing signals from edge sensors has grown explosively. Therefore, there is an urgent need for a lightweight network model that can handle temporal signals. Recurrent Neural Network (RNN) is capable of dealing with temporal tasks. But the training of RNN is usually very difficult and requires extensive computational power, mainly resulting from the problem of exploding or vanishing gradients in recurrent structures. In order to solve this problem, the concept of Reservoir Computing (RC) was proposed.

Since training an RC system only involves training the connection weights in the readout function between the reservoir and output, training cost

FIGURE 6.14 (a) High-precision nvCIM scheme based on MLSS 3D VRRAM to overcome the challenges of applying 2D RRAM to large 3D CNNs. (b) Data flow and architecture of the conventional parallel WL input in situ vector–matrix multiplication (PWIVMM) scheme for 1bIN–1bPW vector–matrix multiplication operation (white lines) and the ADINWM scheme for 8bIN–8bPW VMM operation. (c) Implementation framework of the conventional PWIVMM scheme and proposed an anti-drift multibit analogue input-weight multiplication (ADINWM) scheme. (d) Bit density of the proposed nvCIM macro.[58,60]

can be significantly reduced compared with conventional RNN approaches. But traditional RC architectures are still complex, which employ several hundreds or thousands of nonlinear reservoir nodes for good performance (Figure 6.15a). To further simplify the network structure and reduce the difficulty of hardware implementation, Appellant L et al. introduced a novel architecture that reduces the usually required large number of elements to a single nonlinear node with a delayed feedback.[68] The nodes in the reservoir are virtual. They proved that delay-dynamical system, even in their simple manifestation, can efficiently process information.

The physical implementation of the delayed feedback-based reservoir computing system has received a lot of attention from researchers. The inherent dynamic properties and nonlinear behaviour of dynamic memristors (Figure 6.15e) make them very suitable for the implementation of the delayed feedback-based reservoir computing systems. The internal short-term memory effects of the dynamic memristors allow the memristor-based reservoir to nonlinearly map temporal inputs into reservoir states, where the projected features can be readily processed by a linear readout function (Figure 6.15b). Chao Du et al. constructed reservoirs based on a 32 × 32 array of WO_x memristors, accomplished the task of digital image recognition and handwritten digit recognition, and solved a second-order nonlinear dynamic task.[69] This work is important for the implementation of the delayed feedback reservoir based on memristors. Subsequently, John Moon et al. built reservoirs based on the same WO_x memristors which nevertheless can perform more complex tasks: spoken-digit recognition and chaotic system forecasting.[70]

Zhong et al. used $Ti/TiO_x/TaO_y/Pt$ (50 nm/16 nm/30 nm/50 nm) vertical stacking structure to build a reservoir. In this work, the author considered the impact of masks on the RC system and found that there exist corresponding optimal masks when the RC system executes different tasks.[71] These works all prove that the delayed feedback reservoir is computationally efficient and more friendly to hardware implementation. However, driven by the rapid development of the Internet of Things, high-density and low-power consumption requirements are placed on computing networks. Delayed feedback reservoirs based on planar structures face problems such as a large number of devices and large chip area. To address these issues, increasing the density of devices in 3D architectures is a practical approach.

We constructed 3D reservoirs based on 3D memristors with the advantages of high density and high area efficiency. Because the cycle-to-cycle (C2C) variation of memristors brings about the error source during hardware implementation and degrades the performance of the RC system, we design the parallel memristor configuration. We found that the parallel configuration greatly reduced the change of peak current (Figure 6.15f), and the distribution

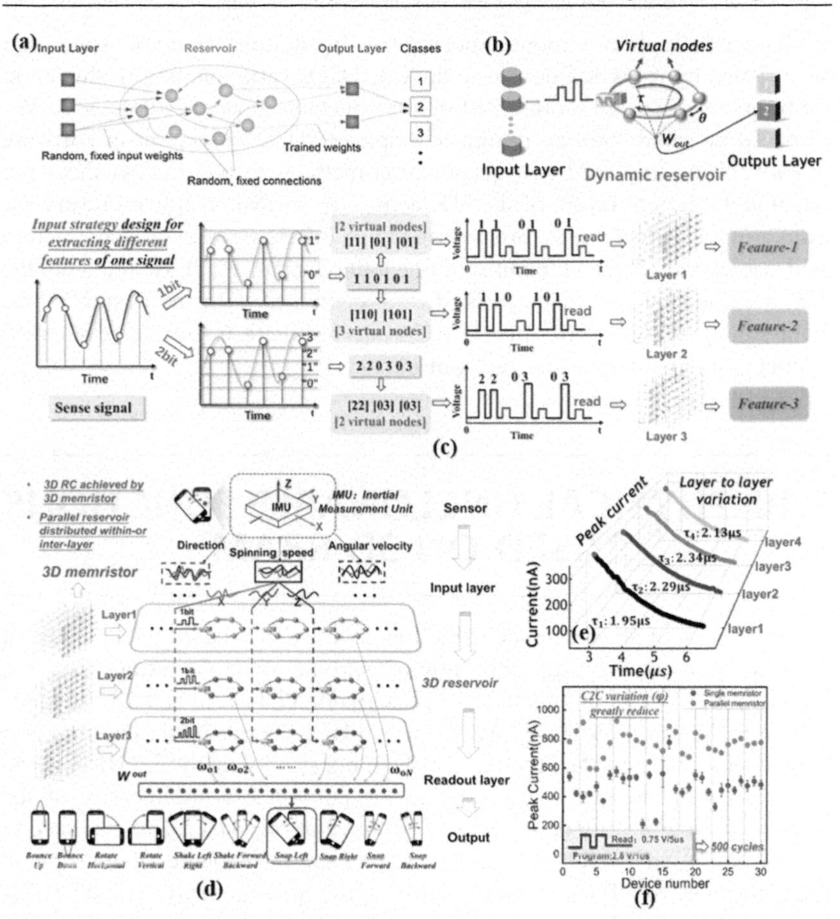

FIGURE 6.15 (a) Classical RC scheme. The input is coupled into the reservoir via a randomly connected input layer to the N nodes in the reservoir. The connections between reservoir nodes are randomly chosen and kept fixed. The reservoir's transient dynamical response is read out by an output layer.[68] (b) Memristor-based delayed feedback reservoir system. (c) Input strategy design. (d) The gesture recognition task executed by the 3D RC system. (e) The dynamic characteristics of memristor in different layers. (f) Hardware optimisation by structure design for the reservoirs. By using the paralleled design, the peak currents are improved and the C2C variation of peak current is well controlled. 6.15(b–f) © [2022] IEEE. Reprinted, with permission, from [Wenxuan Sun, 3D Reservoir Computing with High Area Efficiency (5.12 TOPS/mm^2) Implemented by 3D Dynamic Memristor Array for Temporal Signal Processing, IEEE Proceedings, and Jun. 12, 2022].

of characteristic time is more concentrated. In addition, reservoir states can be enriched by using the device-to-device (D2D) variation, and in this case, the reservoir states are represented by the collective states of all devices. We proposed an input strategy design to improve "D2D" variation in software (Figure 6.15c), and used two quantisation methods to pre-process the input signal and different layers of the 3D memristors array constructed reservoirs with different virtual nodes. So, we used three different reservoir construction modes to extract different features of one unified input signal. By reducing the C2C variation of the device in hardware, combined with the design of the input strategy in software, the 3D RC system (Figure 6.15d) achieves 90% accuracy in dynamic gesture recognition.

6.10 PHYSICAL UNCLONABLE FUNCTIONS BASED ON 3D RRAM

The realisation of physical unclonable function (PUF) is based on the inherent random variation during the microelectronic manufacturing process. Conventional PUF types include ring oscillator PUF, arbiter PUF, bistable PUF, etc. However, for these CMOS-based PUFs, the potential randomness is limited and can be cracked by machine learning attacks easily. 3D RRAM can solve this problem excellently, due to the controllable nonlinearity of the memristor. Nonlinearity depends heavily on the storage states and is related to process variation, which can serve as an important source of entropy in storage arrays.[72–75] Besides, in digital designs, crossbar arrays can be effectively transformed into linear resistor networks, which greatly simplifies input-output mapping.

Yang et al.[76] proposed and demonstrated a multi-layer 3D vertical RRAM (VRRAM) PUF. The eight-layer 3D VRRAM structure in Figure 6.16a possesses the advantages of the crossbar to provide high area efficiency. Figure 6.16b shows the statistics of I-V curves in the eight-layer 8 × 32 arrays. The devices in each layer exhibit stable and uniform self-selective memory characteristics. Given that one-bit PUF output is generated by comparing two RRAM cells, the PUF cell can achieve $8F^2$/bit and equivalently $1F^2$/bit for eight-layer stacking. Such one-bit PUF output is generated by applying a voltage bias to the two selected RRAM cells, and then comparing the total currents running into the two selected paths, as shown in Figure 6.16c. Figure 6.16d shows the distribution of measured HRS of 1K 3D RRAM cells, which demonstrates good randomness.

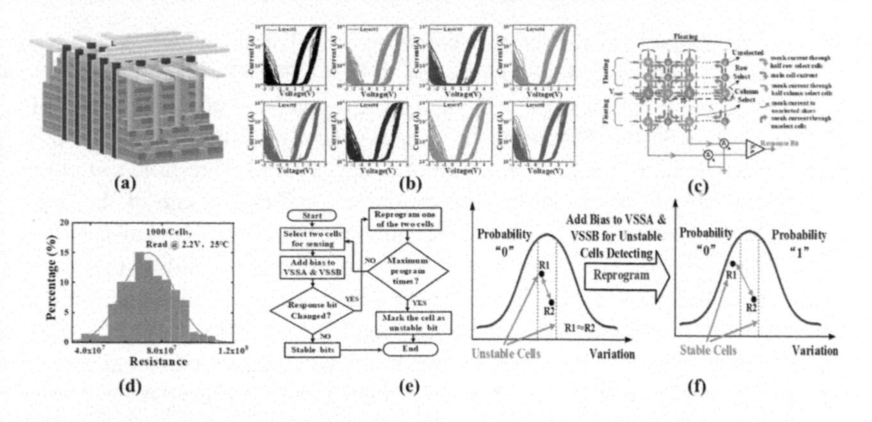

FIGURE 6.16 (a) The schematic of 3D VRRAM architecture. (b) Typical DC I-V curves of eight-layer (Layer 1–Layer 8) RRAM cells. (c) The current path in 3D VRAM during reading. (d) Distribution of the HRS of 1,000 3D RRAM cells. (e) The work flow of the proposed in cell stabilisation scheme. (f) Unstable cells detection and stabilisation when the distribution of mismatch is in the unstable region. © [2020] IEEE. Reprinted, with permission, from [Jianguo Yang, A Machine-Learning-Resistant 3D PUF with Eight-Layer Stacking Vertical RRAM and 0.014% Bit Error Rate Using in-Cell Stabilization Scheme for IoT Security Applications, IEEE Proceedings, and Dec. 12, 2020].

Apart from the randomness in RRAM resistance, enhanced randomness can be obtained from the parasitic IR drop and abundant sneak current paths in 3D RRAM, which means more random sources. However, the 3D stackability has not been fully exploited for further area saving and reliability is still a problem, since two close HRS may induce an unstable PUF output under PVT variations, which could lead to poor BER. To deal with this problem, Jianguo Yang et al. invented an in-cell stabilisation scheme to improve both cost efficiency and reliability. Figure 6.16e shows the workflow for unstable bit detection. If the PUF output is the same when either biasing voltage VSSA or VSSB is applied, a stable bit is identified. Otherwise, an unstable bit is found. Then, if this is an unstable bit, reprogramming, which includes a successive SET and RESET cycle, is employed on one or two of the selected RRAM cells. After that, the detection is performed again to determine if the process will continue. Figure 6.16f shows that the unstable bit issue can be resolved with expanded deviation after reprogramming. The raw BER is 3.08% without stabilisation, and reprogramming three(five) times reduces the BER by >7X (68X) to 0.42% (0.045%). This 3D VRRAM PUF is featured by excellent statistical performance, low cost, and high reliability, showing great potential for security solutions in IoT applications. It can also be scaled up to larger systems composed of many 3D RRAM devices.

Ding et al.[77] put forward a system to leverage the stochasticity of 3D RRAM for stochastic computing in a more unconventional way. The proposed system enables the PUF function in a four-layer 3D threshold switching NbO_X array. Figure 6.17a shows the cross-section TEM image of the four-layer TiN/NbO_X/Pt device. Figure 6.17b illustrates the I-V curves distribution of the NbO_X device in a four-layer 8 × 32 array, which exhibits the stable threshold switching characteristic and discrete leakage current distribution. The schematic mechanism of stochastic origin in the NbO_X device is illustrated in Figure 6.17c. The dynamic threshold switching variations and static leakage current mismatch are utilised for PUF entropy. Figure 6.17d shows the I-V characteristics of NbO_X devices extracted from a four-layer 8 × 32 array. The impact of the intrinsic forming process are manifested as variations in the leakage current and I-V nonlinearity, which are higher at larger biases. The conductance map is measured in a 32 × 32 array (Figure 6.17e) and its distribution (Figure 6.17f) is Gaussian shaped, indicating a rich space of intrinsic variations for cryptographic secret key generation in the 3D NbO_X array. One-bit PUF digitises the leakage current mismatch between the two

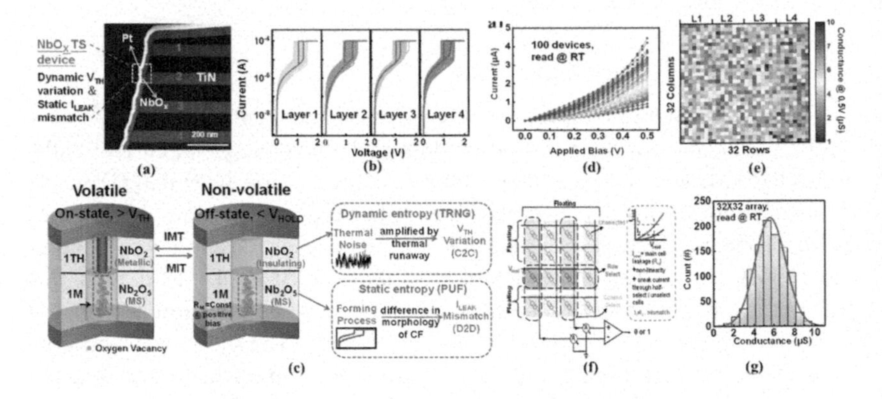

FIGURE 6.17 (a) The cross-section TEM image and (b) typical DC I-V characteristics of four-layer NbO_X cells. (c) Schematics of the stochastic origin in NbO_X. (d) Leakage current characteristics of 100 devices tested in 32 × 32 array. (e) Conductance map of 32 × 32 array for the demonstrated PUF measured with 0.5 V bias at room temperature and (f) its Gaussian-shaped distribution. (g) Working principle of 3D PUF. © [2021] IEEE. Reprinted, with permission, from [Qingting Ding, Unified 0.75 pJ/Bit TRNG and Attack Resilient 2F2/Bit PUF for Robust Hardware Security Solutions with Four-Layer Stacking 3D NbO_x Threshold Switching Array, IEEE Proceedings, and Dec. 11, 2021].

selected cell's bit line under a low voltage bias, while the unselected rows and columns keep floating, and the sensing circuit is illustrated in Figure 6.17g. Ding et al. showed that the 3D NbO_X array is a particularly appropriate device to construct a security platform. Post-silicon data also demonstrated the system with robust security metrics.

H. Nili et al.[78] presented a strong PUF design utilising the nonlinear behaviour of memristors, which was implemented on a two-layer stacked 10 × 10 RRAM array, whose structure is shown in Figure 6.18a. The dielectric layer of RRAM uses Al_2O_3/TiO_{2-x}, which is homogeneous and therefore allows precise conductance adjusting of the individual devices in the array. Figure 6.18b shows the nonlinearity factor calculated as a ratio of $|1 - G_0/G(V_B)|$ for all 200 devices, where G_0 is the device conductance at 200 mV. Figure 6.18b indicates that there is a positive correlation between nonlinearity of the device and reading voltage.

Figure 6.19 shows how the encryption function is implemented in a crossbar circuit with an array size of $M \times N = 20 \times 10$. A binary output b is calculated by biasing the m selected rows with the voltage V_B and then comparing the currents flowing into the two sets of $n/2$ selected adjacent columns as shown below:

$$b = \begin{cases} 1, I^+ > I^- \\ 0, I^+ < I^- \end{cases}, \quad I^\pm = V_B \sum_{j \in S_C^\pm} \sum_{i \in S_R} G_{ij}(V_B) \tag{6.1}$$

(a)

(b)

FIGURE 6.18 (a) The structure of 3D RRAM crossbar array. (b) Nonlinearity characteristic of 3D RRAM device.[78] Reprinted/adapted by permission from [Springer Nature Customer Service Centre GmbH]: [Nature Electronics, Springer Nature] [Hardware-intrinsic security primitives enabled by analogue state and nonlinear conductance variations in integrated memristors] by [Hussein Nili et al.] [COPYRIGHT] (2018).

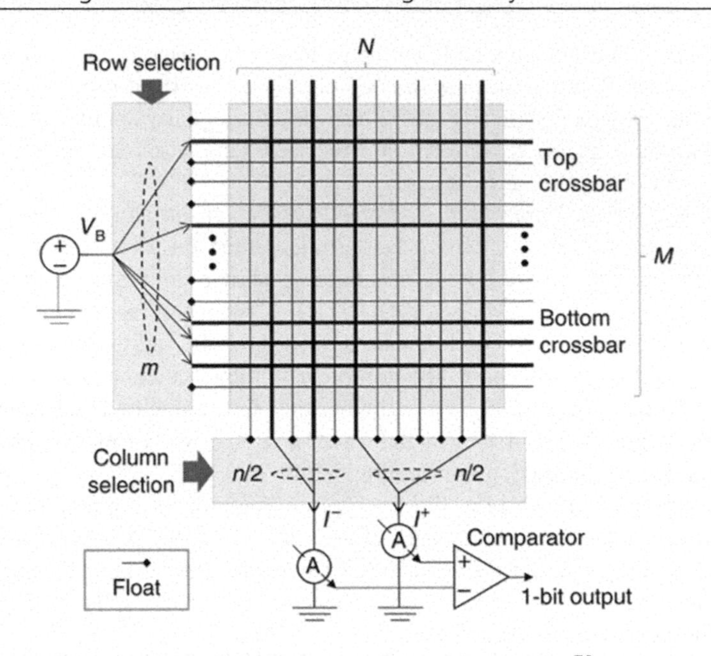

FIGURE 6.19 3D RRAM-based basic building block for PUF.[78]

where S_R is a set of the selected rows, S_C^+ and S_C^- are sets of the selected columns, respectively, and I^+ and I^- are their respective currents. The remaining rows and columns in the array are floating. Using such a scheme, the maximum number of reachable CRP combinations can be calculated by:

$$C_{max} = \binom{M}{m} \times \binom{N}{n} \tag{6.2}$$

It can be observed from formula (6.2) that using this PUF scheme can expand a CRP space, which is much larger than that of the traditional PUF. To enhance the randomness contribution of the device's I-V characteristic as well as improve device reliability, this PUF scheme requires a write-verify algorithm to tune the conductance of the crossbar array to a specific pre-computed value. Meanwhile, the nonlinearity of the device is further exploited by applying different reading voltages to the array.

In general, the inherent nonlinearity of memristors offers greater randomness when serving as the entropy source of a PUF design, and the 3D-stacked RRAM structure provides further enhanced stochasticity as well as higher area efficiency.

These works display the feasibility and reliability of 3D RRAM-based PUF implementation, which could become a good baseline for future study in this field.

REFERENCES, BIBLIOGRAPHY, OR WORKS CITED

[1] A. Mehonic and A.J. Kenyon, "Brain-inspired computing needs a master plan," *Nature*, vol. 604, pp. 255–260, 2022.

[2] K. Roy, A. Jaiswal, and P. Panda, "Towards spike-based machine intelligence with neuromorphic computing," *Nature*, vol. 575, pp. 607–617, 2019.

[3] T. Shi, R. Wang, Z. Wu, Y. Sun, J. An, and Q. Liu, "A review of resistive switching devices: Performance improvement, characterization, and applications," *Small Structures*, vol. 2, p. 2000109, 2021.

[4] S.B. Furber, F. Galluppi, S. Temple, and L.A. Plana, "The SpiNNaker project," *Proceedings of the IEEE*, vol. 102, pp. 652–665, 2014.

[5] L.F. Abbott and W.G. Regehr, "Synaptic computation," *Nature*, vol. 431, pp. 796–803, 2004.

[6] S.H. Jo, T. Chang, I. Ebong, B.B. Bhadviya, P. Mazumder, and W. Lu, "Nanoscale memristor device as synapse in neuromorphic systems," *Nano Letters*, vol. 10, pp. 1297–1301, 2010.

[7] X.M. Zhang, et al., "Emulating short-term and long-term plasticity of bio-synapse based on Cu/a-Si/Pt memristor," *IEEE Electr Device L*, vol. 38, pp. 1208–1211, 2017.

[8] Z. Wang, et al., "Memristors with diffusive dynamics as synaptic emulators for neuromorphic computing," *Nat Mater*, vol. 16, pp. 101–108, 2016.

[9] T. Ohno, T. Hasegawa, T. Tsuruoka, K. Terabe, J.K. Gimzewski, and M. Aono, "Short-term plasticity and long-term potentiation mimicked in single inorganic synapses," *Nat Mater*, vol. 10, pp. 591–595, 2011.

[10] T. Chang, S.-H. Jo, and L.W. Wei, "Short-term memory to long-term memory transition in a nanoscale memristor," *ACS Nano*, vol. 5, pp. 7669–7676, 2011.

[11] Y. Park and J.S. Lee, "Artificial synapses with short- and long-term memory for spiking neural networks based on renewable materials," *ACS Nano*, vol. 11, pp. 8962–8969, 2017.

[12] M.K. Kim and J.S. Lee, "Short-term plasticity and long-term potentiation in artificial biosynapses with diffusive dynamics," *ACS Nano*, vol. 12, pp. 1680–1687, 2018.

[13] Y. Shi, et al., "Electronic synapses made of layered two-dimensional materials," *Nature Electronics*, vol. 1, pp. 458–465, 2018.

[14] W. Xu, et al., "Organometal halide perovskite artificial synapses," *Adv Mater*, vol. 28, pp. 5916–5922, 2016.

[15] D. Purves, G.J. Augustine, D. Fitzpatrick, W.C. Hall, A.-S. LaMantia, and L.E. White *Neuroscience*, 3rd edn. Sinauer Associates (2012).
[16] N. Caporale and Y. Dan, "Spike timing-dependent plasticity: A Hebbian learning rule," *Annual Review of Neuroscience*, vol. 31, pp. 25–46, 2008.
[17] D.E. Feldman, "The spike-timing dependence of plasticity," *Neuron*, vol. 75, pp. 556–571, 2012.
[18] Y. Wu, S. Yu, and H.-S.P. Wong, "Alo_x-based resistive switching device with gradual resistance modulation for neuromorphic device application." In: *2012 IEEE International Memory Workshop (IMW)*). IEEE (2012).
[19] Y. Li, et al., "Activity-dependent synaptic plasticity of a chalcogenide electronic synapse for neuromorphic systems," *Sci Rep*, vol. 4, p. 4906, 2014.
[20] M. Prezioso, F. Merrikh Bayat, B. Hoskins, K. Likharev, and D. Strukov, "Self-adaptive spike-time-dependent plasticity of metal-oxide memristors," *Sci Rep*, vol. 6, p. 21331, 2016.
[21] C. Du, W. Ma, T. Chang, P. Sheridan, and W.D. Lu, "Biorealistic implementation of synaptic functions with oxide memristors through internal ionic dynamics," *Advanced Functional Materials*, vol. 25, pp. 4290–4299, 2015.
[22] R.C. Froemke and Y. Dan, "Spike-timing-dependent synaptic modification induced by natural spike trains," *Nature*, vol. 416, pp. 433–438, 2002.
[23] R. Yang, et al., "Synaptic suppression triplet-STDP learning rule realized in second-order memristors," *Advanced Functional Materials*, vol. 28, p. 1704455, 2017.
[24] Z. Wang, et al., "Toward a generalized Bienenstock-Cooper-Munro rule for spatiotemporal learning via triplet-STDP in memristive devices," *Nat Commun*, vol. 11, p. 1510, 2020.
[25] J. Xiong, et al., "Bienenstock, Cooper, and Munro learning rules realized in second-order memristors with tunable forgetting rate," *Advanced Functional Materials*, vol. 29, p. 1807316, 2019.
[26] A. Kirkwood, M.G. Rioult, and M.F. Bear, "Experience-dependent modification of synaptic plasticity in visual cortex," *Nature*, vol. 381, pp. 526–528, 1996.
[27] R. Jolivet, T.J. Lewis, and W. Gerstner, "Generalized integrate-and-fire models of neuronal activity approximate spike trains of a detailed model to a high degree of accuracy," *J Neurophysiol*, vol. 92, pp. 959–976, 2004.
[28] A.L. Hodgkin and A.F. Huxley, "A quantitative description of membrane current and its application to conduction and excitation in nerve," *The Journal of Physiology*, vol. 117, pp. 500–544, 1952.
[29] M.D. Pickett, G. Medeiros-Ribeiro, and R.S. Williams, "A scalable neuristor built with Mott memristors," *Nature Materials*, vol. 12, pp. 114–117, 2013.
[30] W. Yi, K.K. Tsang, S.K. Lam, X. Bai, J.A. Crowell, and E.A. Flores, "Biological plausibility and stochasticity in scalable VO_2 active memristor neurons," *Nat Commun*, vol. 9, p. 4661, 2018.
[31] S. Kumar, R.S. Williams, and Z. Wang, "Third-order nanocircuit elements for neuromorphic engineering," *Nature*, vol. 585, pp. 518–523, 2020.
[32] J. Lin, et al., "Low-voltage artificial neuron using feedback engineered insulator-to-metal-transition devices." In: *2016 Ieee International Electron Devices Meeting (IEDM)*). IEEE (2016).

[33] X.M. Zhang, et al., "An artificial neuron based on a threshold switching memristor," *IEEE Electr Device L*, vol. 39, pp. 308–311, 2018.
[34] H. Kalita, et al., "Artificial neuron using vertical MoS2/graphene threshold switching memristors," *Sci Rep*, vol. 9, p. 53, 2019.
[35] D. Lee, et al., "Various threshold switching devices for integrate and fire neuron applications," *Advanced Electronic Materials*, vol. 5, p. 1800866, 2019.
[36] Z. Wang, et al., "Fully memristive neural networks for pattern classification with unsupervised learning," *Nature Electronics*, vol. 1, pp. 137–145, 2018.
[37] X. Zhang, et al., "Hybrid memristor-CMOS neurons for in-situ learning in fully hardware memristive spiking neural networks," *Science Bulletin*, vol. 66, pp. 1624–1633, 2021.
[38] M. Prezioso, et al., "Spike-timing-dependent plasticity learning of coincidence detection with passively integrated memristive circuits," *Nat Commun*, vol. 9, p. 5311, 2018.
[39] V. Milo, et al., "Demonstration of hybrid CMOS/RRAM neural networks with spike time/rate-dependent plasticity." In: *2016 IEEE International Electron Devices Meeting*). IEEE (2016).
[40] S. Ambrogio, et al., "Neuromorphic learning and recognition with one-transistor-one-resistor synapses and bistable metal oxide RRAM," *IEEE T Electron Dev*, vol. 63, pp. 1508–1515, 2016.
[41] W. Wang, et al., "Learning of spatiotemporal patterns in a spiking neural network with resistive switching synapses," *Science Advances*, vol. 4, p. 8, 2018.
[42] Q. Duan, et al., "Spiking neurons with spatiotemporal dynamics and gain modulation for monolithically integrated memristive neural networks," *Nat Commun*, vol. 11, p. 3399, 2020.
[43] X. Zhang, et al., "Fully memristive SNNs with temporal coding for fast and low-power edge computing." In: *2020 IEEE International Electron Devices Meeting*) (2020).
[44] X. Zhang, et al., "Experimental demonstration of conversion-based SNNs with 1T1R Mott neurons for neuromorphic inference." In: *2019 IEEE International Electron Devices Meeting*) (2019).
[45] M. Pfeiffer and T. Pfeil, "Deep learning with spiking neurons: Opportunities and challenges," *Front Neurosci*, vol. 12, p. 774, 2018.
[46] D. Zambrano and S.M. Bohte, "Fast and efficient asynchronous neural computation with adapting spiking neural networks." *arXiv preprint arXiv:160902053* (2016).
[47] R. Midya, et al., "Artificial neural network (ANN) to Spiking Neural Network (SNN) converters based on diffusive memristors," *Advanced Electronic Materials*, vol. 5, p. 1900060, 2019.
[48] Y. Kim, et al., "A bioinspired flexible organic artificial afferent nerve," *Science*, vol. 360, p. 998, 2018.
[49] X. Zhang, et al., "An artificial spiking afferent nerve based on Mott memristors for neurorobotics," *Nature Communications*, vol. 11, p. 51, 2020.
[50] J. Zhu, et al., "A heterogeneously integrated spiking neuron array for multimode-fused perception and object classification," *Adv Mater*, vol. 34, p. e2200481, 2022.

[51] I. Wang, et al., "3D synaptic architecture with ultralow sub-10 fJ energy per spike for neuromorphic computation." In: *2014 IEEE International Electron Devices Meeting*) (2014).

[52] I. Wang, et al., "3D Ta/TaO_x/TiO_2/Ti synaptic array and linearity tuning of weight update for hardware neural network applications," *Nanotechnology*, vol. 27, p. 365204, 2016.

[53] Z. Yao, et al., "Simultaneous implementation of resistive switching and rectifying effects in a metal-organic framework with switched hydrogen bond pathway," *Sci Adv*, vol. 5, p. eaaw4515, 2019.

[54] Y.Y. Zhao, et al., "All-inorganic ionic polymer-based memristor for high-performance and flexible artificial synapse," *Adv Funct Mater*, vol. 30, p. 2004245, 2020.

[55] M. Sivan, et al., "All WSe2 1T1R resistive ram cell for future monolithic 3D embedded memory integration," *Nat Commun*, vol. 10, p. 5201, 2019.

[56] G. Ding, et al., "2D metal-organic framework nanosheets with time-dependent and multilevel memristive switching," *Adv Funct Mater*, vol. 29, p. 1806637, 2019.

[57] T.Y. Wang, et al., "Flexible 3D memristor array for binary storage and multi-states neuromorphic computing applications," *InfoMat.*, vol. 3, pp. 212–221, 2020.

[58] B. Kim, et al., "Leveraging 3D vertical RRAM to developing neuromorphic architecture for pattern classification." *2020 IEEE Computer Society Annual Symposium on VLSI* (ISVLSI).

[59] P.-Y. Chen, Z. Li, and S. Yu, "Design tradeoffs of vertical rram-based 3-D cross-point array," *IEEE Transactions on Very Large Scale Integration (VLSI) Systems*, vol. 24, no. 12, pp. 3460–3467, 2016.

[60] H. Li, et al., "Fourlayer 3D vertical rram integrated with finfet as a versatile computing unit for brain-inspired cognitive information processing." In: *IEEE Symposium on VLSI Technology* (2016), pp. 1–2.

[61] Z. Li, P.-Y. Chen, H. Xu, and S. Yu, "Design of ternary neural network with 3-D vertical rram array," *IEEE Transactions on Electron Devices*, vol. 64, no. 6, pp. 2721–2727, 2017.

[62] S. Choi, W. Sun, and H. Shin, "Analysis of read margin and write power consumption of a 3-d vertical RRAM (VRRAM) crossbar array," *IEEE Journal of the Electron Devices Society*, vol. 6, pp. 1192–1196, 2018.

[63] C. Xue, et al., "A 22 nm 2Mb ReRAM compute-in-memory macro with 121-28TOPS/W for multibit MAC computing for tiny AI edge devices," *ISSCC*, pp. 244–246, Feb. 2020.

[64] Q. Liu, et al., "A fully integrated analog ReRAM based 78.4TOPS/W compute-in-memory chip with fully parallel MAC computing," *ISSCC*, pp. 500–502, Feb. 2020.

[65] C. Xue, et al., "A 1Mb multibit ReRAM computing-in-memory macro with 14.6ns parallel MAC computing time for CNN based AI edge processors," *ISSCC*, pp. 388–390, Feb. 2019.

[66] W. Chen, et al., "A 65nm 1Mb nonvolatile computing-in-memory ReRAM macro with sub-16ns multiply-and-accumulate for binary DNN AI edge processors," *ISSCC*, pp. 494–496, Feb. 2018.

[67] Q. Huo, et al., "Demonstration of 3D convolution kernel function based on 8-layer 3D vertical resistive random access memory," *IEEE Electron Device Letter*, vol. 41, no. 3, pp. 497–500, 2020.
[68] L. Appeltant, et al., "Information processing using a single dynamical node as a complex system[J]," *Nature Communications*, vol. 2, no. 1, pp. 1–6, 2011.
[69] C. Du, et al., "Reservoir computing using dynamic memristors for temporal information processing[J]," *Nature Communications*, vol. 8, no. 1, pp. 1–10, 2017.
[70] J. Moon, et al., "Temporal data classification and forecasting using a memristor-based reservoir computing system[J]," *Nature Electronics*, vol. 2, no. 10, pp. 480–487, 2019.
[71] Y. Zhong, et al., "Dynamic memristor-based reservoir computing for high-efficiency temporal signal processing[J]," *Nature Communications*, vol. 12, no. 1, pp. 1–9, 2021.
[72] A. Chen, "Comprehensive assessment of RRAM-based PUF for hardware security applications." In: *2015 IEEE International Electron Devices Meeting (IEDM)*. IEEE (2015).
[73] A. Chen, "Utilizing the variability of resistive random access memory to implement reconfigurable physical unclonable functions," *IEEE Electron Device Letters*, vol. 36, no. 2, pp. 138–140, 2014.
[74] L. Gao, et al., "Physical unclonable function exploiting sneak paths in a resistive cross-point array," *IEEE Transactions on Electron Devices*, vol. 63, no. 8, pp. 3109–3115, 2016.
[75] Y. Pang, et al., "A novel PUF against machine learning attack: Implementation on a 16 Mb RRAM chip." In: *2017 IEEE International Electron Devices Meeting (IEDM)*. IEEE (2017).
[76] J. Yang, et al., "A machine-learning-resistant 3D PUF with 8-layer stacking vertical RRAM and 0.014% bit error rate using in-cell stabilization scheme for IoT security applications." In: *2020 IEEE International Electron Devices Meeting (IEDM)*. IEEE (2020).
[77] Q. Ding, et al., "Unified 0.75 pJ/bit TRNG and attack resilient 2F2/bit PUF for robust hardware security solutions with 4-layer stacking 3D NbOx threshold switching array." *2021 IEEE International Electron Devices Meeting (IEDM)*. IEEE (2021): 39.2.1–39.2.4.
[78] H. Nili, et al., "Hardware-intrinsic security primitives enabled by analog state and nonlinear conductance variations in integrated memristors," *Nature Electronics*, vol. 1, no. 3, pp. 197–202, 2018.